Appli

Applied Science Review™

Biochemistry

Beverly A. Lyman, PhD
Associate Professor
Departments of Clinical Laboratory Sciences
and Pharmaceutical Sciences
University of Tennessee
Memphis

Springhouse Corporation
Springhouse, Pennsylvania

Staff

EXECUTIVE DIRECTOR, EDITORIAL
Stanley Loeb

PUBLISHER, TRADE AND TEXTBOOKS
Minnie B. Rose, RN, BSN, MEd

ART DIRECTOR
John Hubbard

CLINICAL CONSULTANT
Maryann Foley, RN, BSN

EDITORS
Diane Labus, David Moreau, Paula Bakule

COPY EDITORS
Diane M. Armento, Mary Hohenhaus Hardy, Pamela Wingrod

DESIGNERS
Stephanie Peters (associate art director), Matie Patterson (senior designer)

COVER ILLUSTRATION
Scott Thorn Barrows

ILLUSTRATORS
Robert Neumann, Stellarvisions, Mary Stangl

TYPOGRAPHY
David Kosten (director), Diane Paluba (manager), Elizabeth Bergman, Joyce Rossi Biletz, Phyllis Marron, Robin Mayer, Valerie L. Rosenberger

MANUFACTURING
Deborah Meiris (director), Anna Brindisi, Kate Davis, T.A. Landis

EDITORIAL ASSISTANTS
Caroline Lemoine, Louise Quinn, Betsy K. Snyder

©1994 by Springhouse Corporation, 1111 Bethlehem Pike, P.O. Box 908, Springhouse, PA 19477-0908. All rights reserved. Reproduction in whole or part by any means whatsoever without written permission of the publisher is prohibited by law. Authorization to photocopy items for internal or personal use, or the internal or personal use of specific clients, is granted by Springhouse Corporation for users registered with the Copyright Clearance Center (CCC) Transactional Reporting Service, provided the base fee of $00.00 per copy plus $.75 per page is paid directly to CCC, 27 Congress St., Salem, MA 01970. For those organizations that have been granted a license by CCC, a separate system of payment has been arranged. The fee code for users of the Transactional Reporting Service is 0874345723/94 $00.00 + $.75.
Printed in the United States of America.
ASR8-010793

Library of Congress Cataloging-in-Publication Data
Lyman, Beverly A.
    Biochemistry / Beverly A. Lyman.
      p. cm. — (Applied science review)
    Includes bibliographical references and index.
    1. Biochemistry. I. Title. II. Series.
QD415.L95 1994
574.19'2—dc20
ISBN 0-87434-572-3   93-17656
                                                CIP

Contents

Advisory Board and Reviewers vi

Dedication vii

Preface viii

1. **Basics of Biochemistry** 1

2. **Protein Composition and Structure** 21

3. **Bioenergetics** 36

4. **Protein Function and Metabolism** 55

5. **Carbohydrate Structure, Function, and Metabolism** 76

6. **Lipid Structure, Function, and Metabolism** 93

7. **Transmission of Genetic Information** 106

8. **Gene Regulation and Analysis** 128

Appendix: Glossary 143

Selected References 148

Index 149

Advisory Board

Leonard V. Crowley, MD
  Pathologist
  Riverside Medical Center
  Minneapolis;
  Visiting Professor
  College of St. Catherine, St. Mary's
    Campus
  Minneapolis;
  Adjunct Professor
  Lakewood Community College
  White Bear Lake, Minn.;
  Clinical Assistant Professor of Laboratory
    Medicine and Pathology
  University of Minnesota Medical School
  Minneapolis

David W. Garrison, PhD
  Associate Professor of Physical Therapy
  College of Allied Health
  University of Oklahoma Health Sciences
    Center
  Oklahoma City

Charlotte A. Johnston, PhD, RRA
  Chairman, Department of Health
    Information Management
  School of Allied Health Sciences
  Medical College of Georgia
  Augusta

Mary Jean Rutherford, MEd, MT(ASCP)SC
  Program Director
  Medical Technology and Medical
    Technicians—AS Programs;
  Assistant Professor in Medical Technology
  Arkansas State University
  College of Nursing and Health
    Professions
  State University

Jay W. Wilborn, CLS, MEd
  Director, MLT-AD Program
  Garland County Community College
  Hot Springs, Ark.

Kenneth Zwolski, RN, MS, MA, EdD
  Associate Professor
  College of New Rochelle
  School of Nursing
  New Rochelle, N.Y.

Reviewers

Jim Peterson, PhD
  Assistant Professor of Chemistry
  University of Alabama
  Tuscaloosa

Ralph A. Jacobson, PhD
  Professor of Biochemistry
  California Polytechnic State University
  San Luis Obispo

Dedication

To Henry Laboda,
my husband and fellow biochemist.

Preface

This book is one in a series designed to help students learn and study scientific concepts and essential information covered in core science subjects. Each book offers a comprehensive overview of a scientific subject as taught at the college or university level and features numerous illustrations and charts to enhance learning and studying. Each chapter includes a list of objectives, a detailed outline covering a course topic, and assorted study activities. A glossary appears at the end of each book; terms that appear in the glossary are highlighted throughout the book in boldface italic type.

Biochemistry, the study of chemical structures and processes that occur in organisms, provides conceptual and factual information on the various topics covered in most biochemistry courses and textbooks and focuses on helping students to understand:
- the basic concepts of biochemistry, including the unique bonding properties of carbon, chemical equilibria, buffering systems, and cellular components
- the composition, structure, and function of protein, carbohydrates, and lipids in biologic systems
- principles of bioenergetics and cellular metabolism
- how genetic information is transmitted
- the control of gene expression
- recombinant DNA technology.

1

Basics of Biochemistry

Objectives

After studying this chapter, the reader should be able to:
- Name and diagram the various functional groups important to biochemistry.
- Differentiate among the four classes of stereoisomers.
- Describe the types of intramolecular and intermolecular interactions and rank them in terms of bond strength.
- Discuss how the chemical structure of water contributes to its roles as a solvent, a regulator of body temperature, and a participant in biochemical reactions.
- State the Henderson-Hasselbalch equation and define its usefulness in solving problems related to acid-base function.
- Explain the properties of the carbonic acid–bicarbonate buffer system that make it the major buffering system of blood.
- Compare prokaryotic and eukaryotic cells; list and define the functions of the cellular structures found in each.

I. Bonding Properties of Carbon

A. General information
1. Biochemistry is the study of life at the molecular level
2. Biochemistry focuses on the compounds found within living organisms, such as enzymes and deoxyribonucleic acid (DNA), and the processes in which they function, such as catalysis and replication
3. Carbon, with four valence electrons (electrons in the outermost shell), can bond to as many as four of the same or different atoms or groups (see *Functional Groups Important in Biochemistry,* page 2)
4. Structures containing carbon may be in a linear, branched, cyclic, or tetrahedral (four-faced pyramid-shaped) form
5. For biochemical reactions to occur, there usually must be an optimal spatial arrangement between the reacting molecules

B. Stereochemistry of biochemical molecules
1. *Stereochemistry* is the study of the interactions of molecules in three dimensions

Basics of Biochemistry

Functional Groups Important in Biochemistry

The following chart shows the important functional groups in biochemistry. The functional group in each compound is boxed. Note that compounds may contain more than one functional group.

GROUP	STRUCTURE		EXAMPLE
hydroxyl	R—OH	CH_3CH_2—$\boxed{OH}$	ethanol
carbonyl	$\underset{-C-}{\overset{O}{\|}}$		
	$R-\underset{\|}{\overset{O}{\|}}-H$	$CH_3-\boxed{\underset{\|}{\overset{O}{\|}}-H}$	acetaldehyde
	$R-\underset{\|}{\overset{O}{\|}}-R'$	$CH_3-\boxed{\underset{\|}{\overset{O}{\|}}-CH_3}$	acetone
carboxyl	$R-\underset{\|}{\overset{O}{\|}}-OH$	$CH_3-\boxed{\underset{\|}{\overset{O}{\|}}-OH}$	acetic acid
ester	$R-\underset{\|}{\overset{O}{\|}}-OR'$	$H_3C-(CH_2)_{14}-\boxed{\underset{\|}{\overset{O}{\|}}-O-CH_2}$ $H_3C-(CH_2)_7-\underset{H}{\overset{H}{C}}=\underset{}{\overset{}{C}}-(CH_2)_7-C-O-\overset{|}{\underset{|}{C}}-H$ $\qquad\qquad\qquad\qquad\qquad O \quad H_2C-O-\overset{O}{\underset{O^-}{\overset{\|}{P}}}-O-CH_2-CH_2-\overset{+}{N}\begin{smallmatrix}CH_3\\CH_3\\CH_3\end{smallmatrix}$	phosphatidyl choline
amino	$R-NH_2$	$HO-\overset{O}{\underset{\|}{C}}-\overset{\boxed{NH_2}}{\underset{\|}{CH}}-CH_3$	alanine
amido	$R-\overset{O}{\underset{\|}{C}}-NH_2$	$H_2N-\overset{O}{\underset{\|}{C}}-NH_2$	urea
thiol	R—SH	$HO-\overset{O}{\underset{\|}{C}}-\overset{NH_2}{\underset{\|}{CH}}-CH_2-\boxed{SH}$	cysteine
disulfide	R—S—S—R'	$HO-\overset{O}{\underset{\|}{C}}-\overset{NH_2}{\underset{\|}{CH}}-CH_2-\boxed{S-S}-CH_2-\overset{NH_2}{\underset{\|}{CH}}-\overset{O}{\underset{\|}{C}}-OH$	cystine

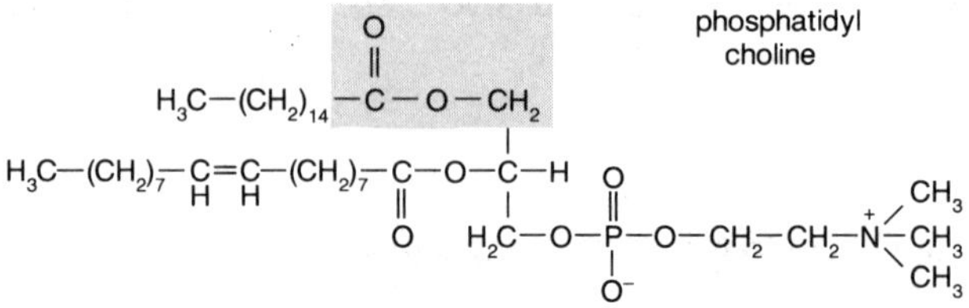

Functional Groups Important in Biochemistry *(continued)*

GROUP	STRUCTURE		EXAMPLE
phospho	$R-O-\overset{\overset{O}{\|\|}}{\underset{\underset{O^-}{\|}}{P}}-O^-$	[phosphate-ribose-adenine structure of AMP]	adenosine monophosphate (AMP)
anhydride	$R-\overset{\overset{O}{\|\|}}{C}-O-\overset{\overset{O}{\|\|}}{\underset{\underset{O^-}{\|}}{P}}-O^-$	$\overset{\overset{O}{\|\|}}{C}-O-\overset{\overset{O}{\|\|}}{\underset{\underset{O^-}{\|}}{P}}-O^-$ $\|$ CHOH $\|$ $CH_2OPO_3^=$	1,3, bis-phosphoglycerate
phenyl	R—⌬	$OH-\overset{\overset{O}{\|\|}}{C}-\overset{\overset{NH_2}{\|}}{CH}-CH_2-$⌬	phenylalanine

2. Carbon atoms in two different compounds may have the same four constituents bound to them, yet differ in the arrangement of constituents; for example, an individual's right and left hands have a palm with five fingers attached, but when the hands are held palm down, the left thumb projects to the right while the right thumb projects to the left
   a. Two compounds having the same substituents and differing only in the arrangement of the substituent groups are **stereoisomers;** an individual's hands could be considered stereoisomers
   b. A primary way of describing stereoisomers is according to whether or not they are mirror images of each other; **diastereoisomers** are stereoisomers that are *not* mirror images of each other; mirror-image stereoisomers are classified by whether or not the mirror-image molecules can be superimposed on each other
   c. **Chiral stereoisomers** are mirror-image stereoisomers that *cannot* be superimposed on each other (an individual's right and left hands are chiral stereoisomers; a right-handed glove cannot be worn on a left hand); an *achiral stereoisomer* is one that *can be* superimposed on its mirror image
   d. A tetrahedral carbon that binds four different substituents is called a *chiral center*
   e. A stereoisomer and its nonsuperimposable mirror image are termed **enantiomers;** an individual's two hands could be considered enantiomers
      (1) Usually, only one of a pair of enantiomers is biologically active; the substituent groups of that molecule form a three-dimensional structure that precisely fits the compound with which it interacts

(2) Enantiomers are usually physically and chemically identical; they differ only by their rotation of polarized light as it passes through a solution of the enantiomer; a device called a polarimeter measures the direction and degree of rotation when viewed looking toward the light source

(3) A (+) enantiomer rotates a plane of polarized light clockwise; a (−) enantiomer rotates a plane of polarized light counterclockwise; a **racemic mixture** contains equal amounts of (+) and (−) enantiomers and is designated (±)

3. The precise arrangement of substituents at a chiral center is its **absolute configuration;** the (+) and (−) designations indicate only the optical activity of the molecule and do not correlate with its absolute configuration; the designation of absolute configuration is based upon several systems, including the Cahn-Ingold-Prelog system and the Fischer projection system

4. The Cahn-Ingold-Prelog system uses the following rules
   a. Assign a priority number to each atom attached *directly to the chiral center;* the higher the atomic number (number of protons in the nucleus), the higher the priority number; the atom with the highest atomic number receives priority number ①; in this scheme ① > ② > ③ > ④
   b. If two atoms attached directly to the chiral center are identical, assign the priority based on the atomic number of the *next* closest atoms; for example, an ethyl group ($-CH_2CH_3$) has higher priority than methyl ($-CH_3$) because the ethyl carbon is attached to two hydrogens (atomic number = 1) and a carbon (atomic number = 6) and the methyl carbon is attached to three hydrogens (atomic number = 1)
   c. Orient the molecule so that the lowest-ranking substituent (④) points away from you
   d. Depict the three highest-ranking substituents as they appear when the reoriented molecule is viewed
   e. If the order of decreasing precedence of these three substituents is clockwise, the absolute configuration is *R*
   f. If the order of decreasing precedence of these three substituents is counterclockwise, the absolute configuration is *S* (see *How to Assign the R or S Configuration to a Chiral Center*)

5. The Fischer projection system uses the following conventions to describe the configuration of groups about a chiral center:
   a. The configuration of groups around the center of any chiral molecule is always related to the configuration of the chiral carbon in glyceraldehyde, a three-carbon carbohydrate with one chiral center; the chiral carbon is indicated by an asterisk

$$\begin{array}{c} H-C=O \\ | \\ H-C^*-OH \\ | \\ CH_2OH \end{array}$$
D-Glyceraldehyde

$$\begin{array}{c} H-C=O \\ | \\ HO-C^*-H \\ | \\ CH_2OH \end{array}$$
L-Glyceraldehyde

How to Assign the *R* or *S* Configuration to a Chiral Center

Assign a priority sequence to the four substituents based upon atomic number. Orient the molecule so that the lowest priority group is behind the plane of the paper; imagine a circular arrow moving from the group with highest priority (①) to the group with next priority (②). If the arrow's path is clockwise, the molecule has the *R* configuration. If the arrow's path is counterclockwise, the molecule has the *S* configuration.

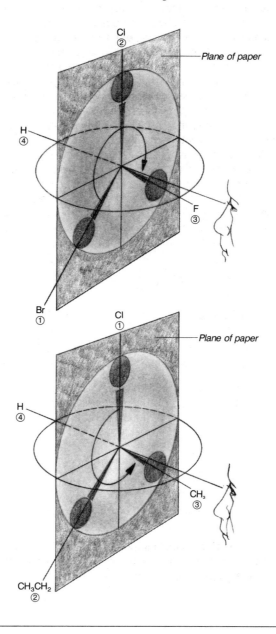

b. The longest carbon chain is drawn vertically; the most oxidized carbon is number 1 and appears at the top
c. Horizontal bonds extend above the plane of the paper (toward the viewer), and vertical bonds extend behind the plane of the paper (away from the viewer)
d. The designations D and L assign the absolute configuration of the chiral center, based on the orientation of the hydroxyl group at the chiral center of glyceraldehyde; the D and L designations do not predict the direction that the enantiomer rotates a plane of polarized light
  (1) The D-stereoisomer is one in which the hydroxyl group on the chiral carbon of glyceraldehyde is written to the right and hydrogen is written to the left in the Fischer projection formula
  (2) The L-stereoisomer is one in which the hydroxyl group on the chiral carbon of glyceraldehyde is written to the left and hydrogen is written to the right in the Fischer projection formula
e. For compounds with more than one chiral center, their absolute configuration is assigned based on whether their highest-numbered chiral center is the same as either D- or L-glyceraldehyde
6. Haworth projection formulas are a conventional way of depicting the cyclic structure of sugars; in this system, the hydroxyl group at the anomeric carbon (the new asymmetric center that forms when a compound is in ring form) is drawn below the plane of the ring (called the α form) or above the plane of the ring (called the β form)

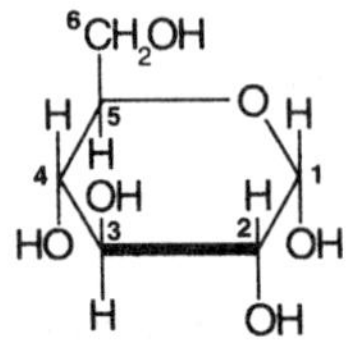

α-D-Glucose
(α-D-Glucopyranose)

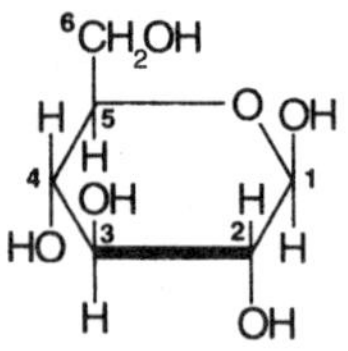

β-D-Glucose
(β-D-Glucopyranose)

7. Ring molecules or molecules with carbon-carbon double bonds can have identical substituents on the same side (labeled *cis* stereoisomers) or on opposite sides of the ring or double bond (labeled *trans* stereoisomers)

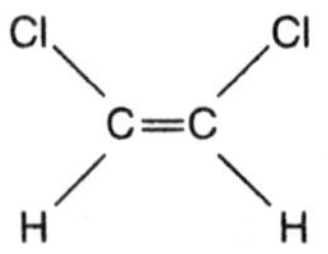

cis 1,2-Dichloroethylene

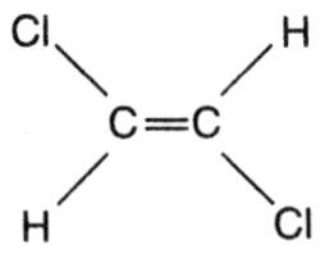

trans 1,2-Dichloroethylene

C. Bond characteristics of biochemical molecules
1. Biochemical reactions involve breaking existing bonds and forming new bonds, either to rearrange structures in the same molecule or to form new molecules

2. Forces of attraction or repulsion between molecules (intermolecular forces) and forces of attraction or repulsion between the atoms comprising a molecule (intramolecular forces) determine the ease with which bonds are broken or formed in chemical reactions
3. Weak bonds, although individually less stable than strong bonds and therefore requiring less energy to break, may collectively constitute very strong forces
    a. The weakest interactions, **van der Waals forces,** are nonspecific attractive forces arising when molecules are an optimal distance apart or nonspecific repulsive forces arising when molecules are too close together
        (1) Van der Waals forces result from mutually induced **dipole** interactions between molecules
        (2) Although individually weak, van der Waals forces become significant when they affect a large number of molecules simultaneously
    b. **Hydrogen bonds** are forces that arise when hydrogen atoms in a molecule form additional bonds with neighboring electronegative atoms, such as oxygen or nitrogen; in many molecules, hydrogen has a partial positive charge and therefore attracts electronegative atoms
        (1) Hydrogen bonds are directional; they are strongest when all three atoms (hydrogen plus the two other atoms in the hydrogen bond) are linear; however hydrogen bonds also can form when the three atoms are at an angle to each other
        (2) Hydrogen bonds can occur within a molecule (intramolecular hydrogen bonding) or between different molecules (intermolecular hydrogen bonding)
    c. Molecules are also characterized as *hydrophilic* (attracted to water molecules) or *hydrophobic* (repulsed by water molecules)
        (1) **Hydrophobic interactions** occur among **nonpolar** groups in an aqueous environment; the nonpolar groups cluster together to minimize the hydrophobic surface area exposed to surrounding water molecules while allowing the water to form as many hydrogen bonds as possible
        (2) **Amphipathic molecules** contain both **polar** (hydrophilic) and nonpolar (hydrophobic) groups in the same molecule; in an aqueous environment, amphipathic molecules form **micelles** in which hydrophobic groups are shielded from water by hydrophilic groups (see *Schematic Representation of a Micelle in Water,* page 8)
    d. **Ionic bonds** are forces of electrostatic attraction between oppositely charged groups; the distance between the charged groups, the amount of electrical charge on each group, and the medium in which the groups reside influences the strength of ionic bonds
        (1) Ionic bonds can be intramolecular or intermolecular
        (2) In contrast to electrostatic attraction between oppositely charged groups, electrostatic repulsion arises between groups bearing the same charge (positive repulses positive; negative repulses negative)
4. Strong bonds are more stable than weak bonds and therefore require more energy to break

Schematic Representation of a Micelle in Water

A micelle is an aggregate of amphipathic molecules in an aqueous environment; the nonpolar groups (represented by short, crooked lines) are shielded from water by polar groups (represented by circles).

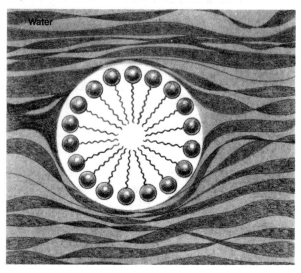

  a. A ***covalent bond*** is a strong bond that forms when two atoms share an electron pair
  b. Atoms in a covalent bond also may share more than one pair of electrons forming, for example, a double bond (two shared electron pairs) or a triple bond (three shared electron pairs)

II. Water

A. General information
1. Water is the major component of cells, comprising 60% to 95% of their total weight
2. Water participates in and regulates biochemical reactions by maintaining the appropriate molar hydrogen ion concentration (represented as $[H^+]$)
3. Water is a good solvent for polar molecules and ionic compounds; water weakens ionic and hydrogen bonds in these compounds by competing with existing atoms for bonding sites; the unique solvent properties of water allow substances to be transported in blood and other fluids and waste products to be solubilized for elimination
4. Water helps maintain constant body temperature in an organism; because of its high specific heat capacity, water absorbs heat well; because of its high heat of vaporization, water gives off heat well

B. The structure of water molecules
1. Water has a dipole: an asymmetric distribution of charge with an area of partial positive charge and an area of partial negative charge (see *Formation of a Hydrogen Bond Between Two Water Molecules*)

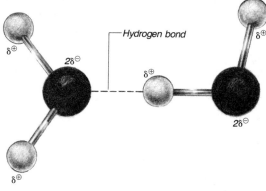

Formation of a Hydrogen Bond Between Two Water Molecules

The $H^{\delta+}$ in one water molecule orients itself toward the $O^{2\delta-}$ in another water molecule, allowing hydrogen bonds to form.

☐ Hydrogen
■ Oxygen

    a. The oxygen (O) atom in a water molecule attracts an electron from each of its two attached hydrogen (H) atoms, thus causing O to have a partial negative charge, represented as $O^{2\delta-}$

    b. The H atoms have a partial positive charge, represented as $H^{\delta+}$

    c. The positively charged $H^{\delta+}$ in one water molecule orients itself toward the negatively charged $O^{2\delta-}$ in another water molecule, allowing hydrogen bonds to form

2. A water molecule can potentially bond with four other water molecules in a very ordered arrangement known as ice

3. Liquid water, while not as ordered as ice, is still an ordered structure with an average of 3.4 hydrogen bonds per molecule

C. The dissociation of water

1. Upon dissociation, a molecule of water forms protons (H^+) and hydroxyl ions (OH^-) in the reaction:

$$H_2O \rightleftarrows H^+ + OH^-$$

2. At equilibrium, the concentrations of [H^+] and [OH^-] versus the concentration of water [H_2O] can be determined experimentally and given a numerical value called the equilibrium constant (K_{eq}) for water

$$K_{eq} = \frac{[H^+][OH^-]}{[H_2O]}$$

    a. The K_{eq} for water has been determined to be 1.8×10^{-16} molar (M)

    b. This very small K_{eq} (meaning that the value of the numerator is very small compared to the value of the denominator) indicates that water normally dissociates to only a very small extent and *most water is in the undissociated form*

3. The concentration of pure water (55.5 M) is calculated using the formula (molecular weight) × (molarity) = grams/liter, where the molecular weight of water = 18 and the weight of 1 liter of water = 1,000 g
4. Substituting values of 1.8×10^{-16} M for K_{eq} and 55.5 M for $[H_2O]$ in the equation above, one can calculate the molar concentration of H^+ and OH^- (represented as $[H^+]$ and $[OH^-]$, respectively) at equilibrium as 1×10^{-14} M
5. Thus, in pure water, the concentration of $[H^+]$ is 1×10^{-7} M and the concentration of $[OH^-]$ is 1×10^{-7} M

III. Acids and Bases

A. General information
1. An acid is a molecule that can donate H^+ to another substance; once it has donated H^+, this same molecule is called a conjugate base because it is now able to accept H^+
2. The acidity of a solution is indicated by its *pH*, a logarithmic measure of the molar concentration of hydrogen ions
3. A base is a molecule that can accept H^+ from another substance; once it has accepted H^+, a base is called a conjugate acid because it is now able to donate H^+
4. A *buffer* is a solution composed of both a weak acid and its conjugate base; a buffer resists large changes in pH because it has both ionized and un-ionized forms present that will either accept any added H^+ or H^-; in biochemical systems buffers are important because they help to maintain a constant pH

B. Definition and measurement of pH
1. pH is the negative logarithm (to the base 10) of $[H^+]$, represented as $-\log[H^+]$; equivalently, pH is $\log\left(\dfrac{1}{[H^+]}\right)$
    a. Expressing $[H^+]$ on a logarithmic scale enables one to indicate very large changes of $[H^+]$ in a solution without writing out very small numbers in scientific notation; for example, if $[H^+]$ changed from 1×10^{-7} to 1×10^{-2}, one can simply say the pH changed from 7 to 2 without writing down the entire concentration of H^+ in scientific notation as a power of 10
    b. The pH scale extends from 0 to 14, accommodating a range of $[H^+]$ from 1 M (pH = 0) to 1×10^{-14} M (pH = 14)
    c. A change in one pH unit indicates a 10-fold change in $[H^+]$, a change in two pH units indicates a 100-fold change in $[H^+]$, and so on
2. Solutions are classified as acidic, basic, or neutral depending on their pH
    a. A solution with a pH of 7 is neutral, indicating equivalent amounts of $[H^+]$ and $[OH^-]$; at pH = 7, $[H^+] = 1 \times 10^{-7}$ M and $[OH^-] = 1 \times 10^{-7}$ M
    b. A solution with a pH > 7 is basic (alkaline), indicating some degree of excess $[OH^-]$ rather than excess $[H^+]$

Acids and Bases 11

Calculating the pH of 1 M Acetic Acid

Acetic acid (H_3CCOOH) has a K'_a of 1.8×10^{-5} M. To find the pH of a 1M solution of acetic acid, set up the following equation:

Step 1. $K'_a = \dfrac{[H^+][H_3CCOO^-]}{[H_3CCOOH]}$ Step 2. $1.8 \times 10^{-5}\,M = \dfrac{[H^+][H_3CCOO^-]}{[H_3CCOOH]}$

Use 1 in the denominator of the K'_a equation because the concentration of 1 M acetic acid at equilibrium is actually this:

$$1M - [\text{concentration of dissociated acid}]$$

Because the concentration of dissociated acid is very small (acetic acid is a weak acid), consider the value of [concentration of dissociated acid] to be insignificant and assign the numerator the value of 1M.

Step 3. $1.8 \times 10^{-5}\,M = \dfrac{[H^+][H_3CCOO^-]}{1M}$ Step 4.* $\sqrt{1.8 \times 10^{-5}\,M} = [H^+]$

Step 5. $4.24 \times 10^{-3}\,M = [H^+]$

From the calculation above, the concentration of $[H^+]$ is 0.00424 M. Using the definition of pH, calculate the final pH of 1M acetic acid as 2.37.

Step 6. $pH = -\log[H^+]$ Step 7. $pH = -\log 4.24 \times 10^{-3}$
$pH = 2.37$

*Note: In equation #4, take the square root of 1.8×10^{-5} as the value of $[H^+]$. Remember that every molecule of H_3CCOOH that dissociates gives one molecule each of H^+ and H_3CCOO^-. Therefore, in this solution, $[H^+] = [H_3CCOO^-]$, and $[H^+] \times [H_3CCOO^-] = [H^+]^2$.

---

    c. A solution with a pH < 7 is acidic, indicating some degree of excess $[H^+]$ rather than excess $[OH^-]$

C. **The equilibrium constant of acids**
    1. Acids are classified as either strong or weak, depending upon their degree of dissociation in an aqueous solution
       a. Just as one can write a formula for the equilibrium constant of water, one can write a formula for the equilibrium constant of any unknown acid; identifying the H^+ donated by an unknown acid as H and the conjugate base of the acid as X, one can call the acid HX and write its equilibrium constant in the following manner:

$$K_{eq} = \dfrac{[H^+][X^-]}{[HX]}$$

       b. A strong acid is one that dissociates completely in water into H^+ and its corresponding anion (negatively charged ion); in an aqueous solution, a strong acid exists mostly in its dissociated (ionized) form and has a high K_{eq}
       c. A weak acid does not dissociate completely in water; it exists partially in its undissociated (un-ionized) form and partly in its dissociated (ionized) form and has a low K_{eq}

Basics of Biochemistry

Derivation of the Henderson-Hasselbalch Equation

The Henderson-Hasselbalch equation is derived from the equation for the equilibrium constant of an acid ($HX \rightleftarrows H^+ + X^-$).

Step 1. $K'_a = \dfrac{[H^+][X^-]}{[HX]}$

Solve the equation by dividing through by $[H^+]$ and K'_a

Step 2. $\dfrac{1}{[H^+]} = \dfrac{1}{K'_a} \times \dfrac{[X^-]}{[HX]}$

Take the logarithm of both sides of the equation

Step 3. $\log \dfrac{1}{[H^+]} = \log \dfrac{1}{K'_a} + \log \dfrac{[X^-]}{[HX]}$

Remember that $pH = \log \dfrac{1}{[H^+]}$ and $pK'_a = \log \dfrac{1}{K'_a}$

Step 4. $pH = pK'_a + \log \dfrac{[X^-]}{[HX]}$ or $pH = pK'_a + \log \dfrac{[\text{proton acceptor}]}{[\text{proton donor}]}$

---

2. In measuring the dissociation characteristics of weak acids, we use the term K'_a, called the apparent dissociation constant (also called the ionization constant)
   a. The term "apparent," abbreviated by the prime symbol ('), indicates that the ionization constant was calculated based on experimentally determined concentrations of reactants and products and is not corrected for deviations from ideal behavior; the subscript "a" in K'_a simply indicates acid
   b. K'_a for a given acid is determined in much the same way that the extent of dissociation of water is obtained from its K_{eq}
   c. With the K'_a of a given acid, one can calculate its pH at any given concentration (see *Calculating the pH of 1 M Acetic Acid,* page 11)

D. **The Henderson-Hasselbalch equation**
   1. The Henderson-Hasselbalch equation relates the pH, pK'_a, and the ratio of concentrations of ionized and un-ionized forms of an acid or base; the Henderson-Hasselbalch equation is written in two comparable ways (see *Derivation of the Henderson-Hasselbalch Equation*)

$$pH = pK'_a + \log \dfrac{[\text{proton acceptor}]}{[\text{proton donor}]}$$

or

$$pH = pK'_a + \log \dfrac{[\text{ionized form}]}{[\text{un-ionized form}]}$$

where $pK'_a = -\log K'_a$

Calculation of K'_a for 1 M Acetic Acid

Using the Henderson-Hasselbalch equation, we can calculate the K'_a of a weak acid if we measure its pH. Suppose the pH of 1M acetic acid is 2.37. Calculate the K'_a of the acid.

Recall the relationship between pH and $[H^+]$.

Step 1. $pH = -\log [H^+]$

In the case of 1 M acetic acid, pH is 2.37.

Step 2. $-\log [H^+] = 2.37$

Step 3. $\log [H^+] = -2.37$

Taking the antilog of both sides of the equation, calculate the actual concentration of H^+.

Step 4. $[H^+] = 0.00424$ M

Using the Henderson-Hasselbalch equation, calculate the K'_a for acetic acid.

Step 5. $pH = pK'_a + \log \dfrac{[H_3CCOO^-]}{[H_3CCOOH]}$

In equation (5) above, subsitute the measured pH of 2.37. Assume that the value of $[H_3CCOO^-]$ is 0.00424 M, the same as the value calculated for $[H^+]$. Because acetic acid is a weak acid, assume a 1 M solution of acid exists mostly in the undissociated form; therefore, assign a value of 1M to $[H_3CCOOH]$.

Substituting these known values into (5) above produces the following equation:

Step 6. $2.37 = pK'_a + \log \dfrac{0.00424 \text{ M}}{1 \text{M}}$

Using a calculator or logarithm table, find the value of the logarithm of 0.00424 (-2.37). Substituting this value into equation (6) above, calculate the pK'_a and then the K'_a for acetic acid.

Step 7. $2.37 = pK'_a + (-2.37)$

$2.37 = pK'_a - 2.37$

$4.74 = pK'_a$

antilog $4.74 = K'_a$

$1.8 \times 10^{-5} = K'_a$

2. Knowing any two of the three parameters in the equation above (pH, ionization constant, or ratio of concentration of ionized and un-ionized forms) allows calculation of the third unknown parameter (see *Calculation of K'_a for 1 M Acetic Acid*)

3. Inspection of the Henderson-Hasselbalch equation reveals that when there are equal concentrations of ionized and un-ionized forms, the Henderson-Hasselbalch equation becomes quite simple:

pH = pK'_a + log 1 (where [ionized form] = [un-ionized form])
and
pH = pK'_a (because log 1 = 0)

4. The relationship among pH, pK'_a, and the ratio of ionized and un-ionized forms is illustrated by a curve plotting pH versus equivalents of base added in the titration of a weak acid—in this case, acetic acid
   a. A weak acid, by definition, exists mainly in the un-ionized form; relatively little is in the ionized form
   b. Adding a small amount of a strong base to a weak acid causes H^+ to dissociate from the acid and join with OH^- from the base to form water; pH increases because the concentration of H^+ decreases
   c. Gradual addition of more base eventually produces a point at which the concentration of the ionized form of the acid equals the concentration of the un-ionized form, as indicated by the Henderson-Hasselbalch equation; at this point in the titration curve, pH = pK'_a

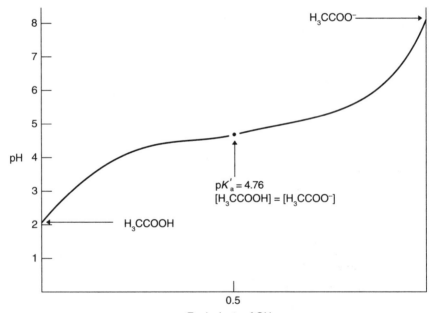

   d. Because equivalent amounts of both ionized and un-ionized forms are present at the pK'_a, addition or removal of acid or base is compensated by the appropriate form either accepting or donating H^+
   e. An acid/conjugate base system that resists change in pH around the pK'_a of the acid is said to be buffered; a *buffer* is a solution of a weak acid and its conjugate base that resists large changes in pH within a certain pH range associated with the pK'_a of the acid

f. In the titration curve shown, addition of even more base gradually brings about a point where the ionized form of the acid predominates, because all H^+ from the un-ionized form have dissociated to join OH^- and form water; at this point, there is a large increase in pH upon further addition of base

IV. Physiologic Buffers

A. General information
1. Blood and other body fluids must be buffered to maintain the optimal intracellular and extracellular pH; optimal pH is necessary to maintain the molecular structure and activity of metabolic enzymes and other proteins
2. The normal pH of blood is 7.40 (range 7.35 to 7.45); blood pH lower than 6.8 or higher than 7.8 is incompatible with life
3. Acidosis, a lower than normal blood pH, is caused by either excess acid or a deficiency of alkali (base) in the body
    a. Respiratory acidosis occurs when the lungs are unable to remove excess CO_2, such as when respiratory problems cause hypoventilation
    b. Metabolic acidosis occurs when excess acid is produced in the body, such as when keto acids accumulate in the blood of diabetic individuals
4. Alkalosis, a higher than normal blood pH, is caused by either excess base or a deficiency of acid in the body
    a. Respiratory alkalosis occurs when the lungs remove too much CO_2, such as during hyperventilation
    b. Metabolic alkalosis occurs when excess acid is lost through nonrespiratory mechanisms, such as loss of gastric acid during prolonged vomiting
5. The major buffering system of blood is the carbonic acid–bicarbonate buffer system (H_2CO_3–HCO_3^-)
6. In the body, the kidneys and the lungs are the major organs that maintain physiologic pH

B. The carbonic acid–bicarbonate buffer system
1. The carbonic acid–bicarbonate buffer system has the following components and reacts as indicated:

$$H_2O + CO_2 \rightleftarrows H_2CO_3 \rightleftarrows H^+ + HCO_3^-$$

2. This system maintains a balance between CO_2, the major acid produced by the body via metabolic processes, and H^+, which must be either added or removed to maintain blood pH
    a. Although CO_2 is considered an acid, it is really an acid anhydride (a compound formed when two molecules of acid react with the loss of one molecule of water); CO_2 reacts with water to form the true acid, carbonic acid (H_2CO_3), in the reaction:

$$H_2O + CO_2 \rightleftarrows H_2CO_3$$

16 Basics of Biochemistry

Calculating the Normal Bicarbonate:Carbon Dioxide Ratio in Blood

The Henderson-Hasselbalch equation demonstrates that blood normally has a ratio of bicarbonate to carbon dioxide of about 20:1.

Step 1. $\text{pH} = pK'_a + \log \dfrac{[\text{proton acceptor}]}{[\text{proton donor}]}$

The normal pH of blood is 7.4.

Step 2. $7.4 = pK'_a + \log \dfrac{[\text{proton acceptor}]}{[\text{proton donor}]}$

From experimentation, it is known that the pK'_a of the carbonic acid/bicarbonate buffer system is 6.1

Step 3. $7.4 = 6.1 + \log \dfrac{[\text{proton acceptor}]}{[\text{proton donor}]}$

Remember that CO_2 is the acid anhydride of carbonic acid. Using HCO_3^- as the proton acceptor and CO_2 as the proton donor yields the following equation:

Step 4. $7.4 = 6.1 + \log \dfrac{[HCO_3^-]}{[CO_2]}$

Bring the unknown quantities to one side of the equation.

Step 5. $7.4 - 6.1 = \log \dfrac{[HCO_3^-]}{[CO_2]}$ Step 6. $1.3 = \log \dfrac{[HCO_3^-]}{[CO_2]}$

Remove the log function by taking the antilog of both sides of the equation.

Step 7. $\text{antilog } 1.3 = \dfrac{[HCO_3^-]}{[CO_2]}$ Step 8. $19.95 = \dfrac{[HCO_3^-]}{[CO_2]}$

---

    b. Carbonic acid rapidly dissociates into hydrogen and bicarbonate ions in the reaction:

$$H_2CO_3 \rightleftarrows H^+ + HCO_3^-$$

    c. The carbonic acid component is usually omitted and the equation is written:

$$H_2O + CO_2 \rightleftarrows H^+ + HCO_3^-$$

    d. This reaction is catalyzed by the enzyme carbonic anhydrase

3. The carbonic acid–bicarbonate buffer system is an excellent blood buffer because it is an open system

    a. It is called an open system because CO_2 is released to the ambient air in the process of breathing; CO_2 can be adjusted to maintain body pH by either hypoventilation or hyperventilation, according to the following reaction:

$$CO_2 \text{ (gas in lungs)} \rightleftarrows CO_2 \text{ (dissolved in blood)}$$

b. The carbonic acid–bicarbonate system has a pK'_a of 6.1; blood pH is maintained at 7.4; normally the carbonic acid–bicarbonate system would not be a good buffer for blood because buffers do not function well at a pH more than one unit beyond their pK'_a
c. In the body, however, this system works efficiently as a buffer for blood because any increased CO_2 produced may be removed by hyperventilation (deep, rapid breathing), thus keeping the blood ratio of HCO_3^- to CO_2 within normal range
d. Similarly, if blood pH rises because of excess base, CO_2 is retained in the blood by hypoventilation (slow breathing), thus restoring the normal ratio of HCO_3^- to CO_2

4. The Henderson-Hasselbalch equation can identify the optimal ratio of CO_2 to HCO_3^- required to maintain blood at pH 7.4 (see *Calculating the Normal Bicarbonate:Carbon Dioxide Ratio in Blood*)
   a. To achieve a blood pH of 7.4, the carbonic acid–bicarbonate buffer system must maintain a blood concentration of HCO_3^- at 20 times the blood concentration of CO_2
   b. Using the Henderson-Hasselbalch equation along with the experimentally determined value of pK'_a of 6.1 for the carbonic acid–bicarbonate buffer system, one can calculate the blood pH of an individual having the abnormal ratio of HCO_3^- to CO_2 of 10:1 rather than the normal 20:1

$$pH = pK'_a + \log \frac{[HCO_3^-]}{[CO_2]}$$

$$pH = 6.1 + \log \frac{10}{1}$$

$$pH = 6.1 + 1$$

$$pH = 7.1$$

   c. Solving the equation for pH reveals that the blood pH is more acidic than normal, confirming the observation that a rise in blood CO_2 or a fall in HCO_3^- accompanies a decrease in blood pH

5. The lungs and the kidneys are the major organs that function to maintain blood within its normal pH range in the body
   a. The lungs help maintain normal blood pH by varying the amount of CO_2 extracted from the blood during respiration; hyperventilation reduces blood CO_2 concentration; hypoventilation increases blood CO_2 concentration
   b. The kidneys help maintain normal blood pH by controlling the composition of the ultrafiltrate of blood, which becomes urine
      (1) H^+ can react with HCO_3^- resulting in reabsorption of bicarbonate back into the blood
      (2) H^+ can be excreted as part of a buffering system, particularly phosphate buffer
      (3) H^+ can be excreted by reacting with ammonia to form NH_4^+

C. Other important biochemical buffers
1. The free amine groups on amino acids make proteins good buffers; the protein hemoglobin is the second most important blood buffer
2. Both inorganic phosphates (HPO_4^{2-} and $H_2PO_4^-$) and organic phosphate esters are important physiologic buffers inside cells

V. Cell Structure

A. General information
1. The cell functions to replicate itself, to protect itself from harmful environmental conditions, and to maintain a discrete environment where diverse biochemical reactions can occur efficiently
2. Knowledge of both the structures within a cell and the arrangement and communication among cells is important in understanding biochemical processes; subcellular structures performing defined functions within the cell are called *organelles*
3. Cells are classified as either **prokaryotic** or **eukaryotic** depending on their organizational structures
   a. Prokaryotic cells lack a nucleus, lack membrane-bound organelles, and have only one chromosome that consists of double-stranded deoxyribonucleic acid (DNA); examples of prokaryotes are bacteria, rickettsiae, and blue-green algae
   b. Eukaryotic cells contain a nucleus, have membrane-bound organelles, and have more than one chromosome that consists of double-stranded DNA bound to proteins; examples of eukaryotes are fungi, unicellular protozoa, and multicellular organisms (both plants and animals)

B. Components of prokaryotic cells
1. Prokaryotes have a *cell wall* and a *cell membrane;* the cell wall, composed of lipopolysaccharides (lipids and carbohydrates), encloses and protects cellular contents
2. The cell membrane (also called the plasma membrane) is composed of lipoproteins (lipids and proteins) and selectively controls the movement of substances from the external and internal environments; the cell membrane is also the site of many enzymatic biochemical reactions
3. Instead of a well-defined nucleus, prokaryotes have a *nuclear region,* in which double-stranded DNA is concentrated
4. The *cytosol,* the portion of the cell outside the nuclear region, is a soluble intracellular matrix containing ions and proteins and is the site of many enzymatic biochemical reactions, such as glycolysis (see Chapter 3, Bioenergetics)
5. *Ribosomes,* masses of ribonucleic acid (RNA) and protein, are distributed throughout the cytosol and are the sites on which protein synthesis occurs
6. Distributed throughout the cytosol are storage granules of carbohydrates that serve as reserve energy supplies

C. Components of eukaryotic cells

1. The *plasma membrane* of eukaryotes surrounds and regulates cellular contents; it is composed of a bilayer of phospholipids in which proteins are embedded
   a. Plasma membrane proteins aid in cellular recognition and communication and facilitate biochemical reactions; specific binding proteins are **receptors** that, when the correct ligand is attached, initiate further biochemical reactions
   b. Plasma membrane proteins help maintain homeostasis by serving as pumps and channels to allow the selective influx and efflux of ions and molecules
2. The *cytosol,* or cytoplasm, fills the interior space (matrix) of the cell; enclosed by the cell membrane, the cytosol is the soluble portion of the cell that surrounds the membrane-bound organelles
   a. The cytosol contains a high concentration of protein and is the site of many enzyme-catalyzed reactions
   b. Free *ribosomes* (not attached to the endoplasmic reticulum) are located in the cytosol; proteins synthesized on free ribosomes are used within the cell
3. The eukaryotic *nucleus* is the site of DNA synthesis, DNA repair, and RNA synthesis
   a. A *nuclear envelope* encloses the nucleus but contains pores to allow communication between the nucleus and the surrounding cytosol
   b. DNA in the nucleus is complexed with proteins and is arranged in chromosomes
   c. The *nucleolus* is the site within the nucleus at which ribosomes are assembled and RNA processing reactions occur
4. *Mitochondria* are organelles in which carbohydrates, lipids, and proteins are oxidized to produce energy for the cell through a series of reactions called oxidative phosphorylation; much of the extracted energy is stored in the form of adenosine triphosphate (see Chapter 3, Bioenergetics)
   a. Mitochondria are enclosed by an *outer* and *innner mitchondrial membrane;* the inner membrane invaginates into the interior space (the matrix) of the mitochondria in folds called cristae
   b. The mitochondrial matrix contains many enzymes and is therefore the site of many enzyme-catalyzed reactions
   c. Mitochondria are unique in that they have their own DNA and their own protein synthetic apparatus
5. The *endoplasmic reticulum* is a series of membrane-bound channels, called cisternae, that transverse the cell from the nuclear envelope to the plasma membrane
   a. The endoplasmic reticulum acts as a conduit for release of proteins and **hormones** from a cell
   b. Rough endoplasmic reticulum is so named because it contains ribosomes, the sites of protein synthesis, on its surface; proteins synthesized on ribosomes bound to the endoplasmic reticulum are exported from the cell
   c. Smooth endoplasmic reticulum lacks ribosomes; it functions primarily in lipid metabolism and also as a detoxification site for harmful substances

       d. Microsomes are the fragments of endoplasmic reticulum that are recovered after *in vitro* separation from other cellular components
6. The *Golgi apparatus,* a tubular system contiguous with the endoplasmic reticulum, modifies and processes molecules to be exported from the cell or to be used by other organelles
7. *Lysosomes* contain degradative hydrolytic enzymes; as "vacuum cleaners" of the cell, they digest foreign and senescent (old) cellular material
8. *Peroxisomes* degrade molecules that produce the toxin hydrogen peroxide (H_2O_2); however, peroxisomes, through the enzyme catalase, also detoxify hydrogen peroxide directly
9. The *cytoskeleton* is a series of protein microtubules and microfilaments that maintain cell shape, function in intracellular transport of substances, and facilitate cell motility and division
10. Plant cells are eukaryotes with unique additional structures
    a. Outside the plasma membrane, plant cells have a *cell wall* composed of cellulose in which proteins and other carbohydrates are embedded; the cell wall protects, strengthens, and shapes the cell
    b. *Chloroplasts,* double-membraned organelles, convert light energy to chemical energy in the process of **photosynthesis**
    c. *Glyoxysomes,* specialized plant organelles, are sites of the glyoxylate pathway, a series of reactions unique to plants; the glyoxylate pathway enables plants to convert fatty acids to glucose, a biochemical conversion not possible for animals
    d. *Vacuoles,* organelles usually found only in plant cells, store nutrients or waste substances that accummulate faster than the cells can excrete them into the environment; vacuoles tend to increase in number and size as a plant cell ages

Study Activities

1. Identify and name the major functional group or groups in any compound.
2. Compare and contrast the following designations of stereoisomers: (+) and (−), D and L, *R* and *S, cis* and *trans.*
3. List the types of intramolecular and intermolecular interactions that occur in molecules and rank them in terms of their relative strength.
4. Calculate the pH of a solution when either the $[H^+]$ or $[OH^-]$ is given.
5. Using the Henderson-Hasselbalch equation, calculate either the pH, pK'_a, or the ratio of the ionized to un-ionized forms when two of these three parameters are given for an acid. State the conditions under which pH = pK'_a.
6. Use the Henderson-Hasselbalch equation to determine the concentration of CO_2 and HCO_3^- at pH values ranging from 6.8 to 7.8.
7. For each subcellular structure, classify it as belonging to either prokaryotic cells eukaryotic cells, or both.
8. State the major function of each subcellular structure.
9. Define the terms listed in bold-faced type throughout this chapter.

2

Protein Composition and Structure

Objectives

After studying this chapter, the reader should be able to:
- Predict the chemical reactivity of each amino acid based upon its structure.
- Discuss the levels of structural organization in proteins.
- Discuss the changes in ionizable groups of amino acids as a function of changes in pH.
- Describe how a protein's primary structure is determined.
- Describe the physical and chemical techniques for determining the presence, separation, characterization, and structure of proteins.

I. Protein Composition

A. General information
1. *Proteins* are precisely defined sequences of *amino acids (residues)* linked through *peptide bonds* to form *polymers* (hundreds or thousands of molecules combined to produce a single macromolecule)
2. A *peptide* is a molecule formed by linking together a small number of amino acids (from two to a dozen) through peptide bonds; both peptides and proteins usually have a free amine group on one end of the molecule (the N-terminal residue) and a free carboxyl group on the other end of the molecule (the C-terminal residue) (see *A Small Peptide of Six Amino Acids,* page 22)
3. The structure, function, and activity of a protein is determined to a large extent by the number, sequence, and chemical properties of its constituent amino acids
4. All proteins are composed of varying sequences of 20 amino acids called the *common amino acids*
    a. Other amino acids that occur in proteins are modifications of the common amino acids
    b. Modifications include addition of functional groups and bond formation between certain groups within the amino acid
    c. Amino acids containing modifications are labeled *derived* or *modified* amino acids; modifications occur after the amino acids are incorporated into the protein

A Small Peptide of Six Amino Acids

Proteins differ from the simple peptide shown below by having hundreds or thousands of amino acids linked through peptide bonds; note that for both peptides and proteins, the two end functional groups are usually different; one end group is usually an amine group (the N-terminal) and the other is usually a carboxyl group (the C-terminal).

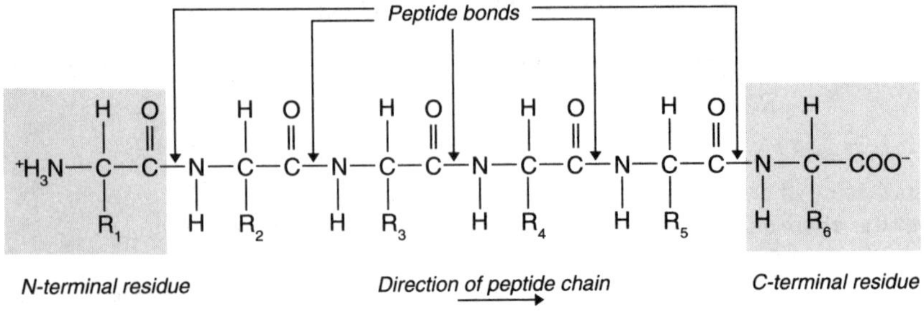

$$^+H_3N-\underset{R_1}{\overset{H}{\underset{|}{C}}}-\overset{O}{\underset{\|}{C}}-\underset{H}{\overset{}{N}}-\underset{R_2}{\overset{H}{\underset{|}{C}}}-\overset{O}{\underset{\|}{C}}-\underset{H}{\overset{}{N}}-\underset{R_3}{\overset{H}{\underset{|}{C}}}-\overset{O}{\underset{\|}{C}}-\underset{H}{\overset{}{N}}-\underset{R_4}{\overset{H}{\underset{|}{C}}}-\overset{O}{\underset{\|}{C}}-\underset{H}{\overset{}{N}}-\underset{R_5}{\overset{H}{\underset{|}{C}}}-\overset{O}{\underset{\|}{C}}-\underset{H}{\overset{}{N}}-\underset{R_6}{\overset{H}{\underset{|}{C}}}-COO^-$$

Peptide bonds are indicated between each residue.

N-terminal residue — Direction of peptide chain → — C-terminal residue

5. The names of the 20 common amino acids are usually written as a three-letter abbreviation with the first letter capitalized or as a single capital letter designation (see *Names and Structures of the 20 Common Amino Acids*)

B. Structure of amino acids

1. Amino acids contain a central carbon (C), called the α carbon, to which four substituents are bound: an amine group ($-NH_2$), a carboxylic acid group ($-COOH$), a hydrogen atom (H), and a group unique to each amino acid called the side chain or R group

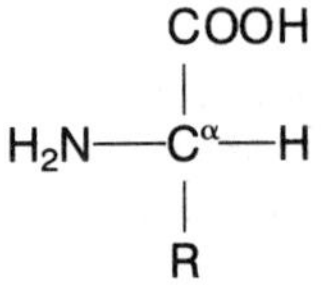

$$H_2N-\underset{R}{\overset{COOH}{\underset{|}{C^\alpha}}}-H$$

   a. Three of the four substituents on the α carbon ($-NH_2$, $-COOH$, and $-H$) are common to all amino acids
   b. The fourth substituent, the side chain (indicated by R), varies and confers distinctive chemical properties to each amino acid
   c. In standard nomenclature for amino acids, the carbons of the R group are labeled with Greek letters, starting with β for the first carbon attached to the α carbon; for example, in glutamic acid (three linear carbons in the R group), the carbons are labeled β, γ, and δ, with the δ carbon being the final carboxyl carbon in this R group
2. All amino acids found in proteins are called α amino acids because the amine group is bonded to the α carbon; however, amino acids with an amine group attached to other carbon atoms have important biochemical functions

Names and Structures of the 20 Common Amino Acids

The charge shown on each amino acid below reflects the charge present at pH = 7.0.

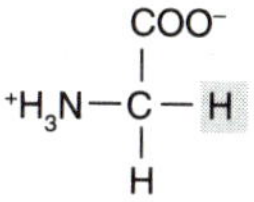

Glycine (Gly, G) NP
Alanine (Ala, A) NP
Valine (Val, V) NP
Leucine (Leu, L) NP
Isoleucine (Ile, I) NP

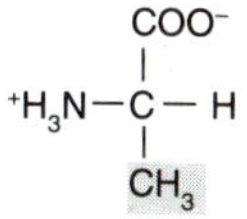

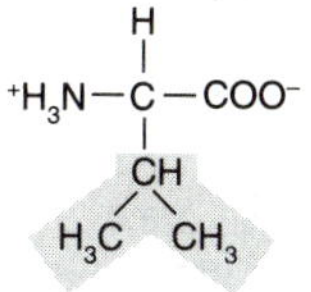

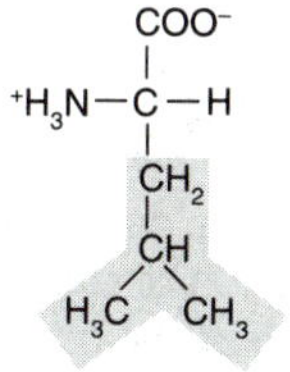

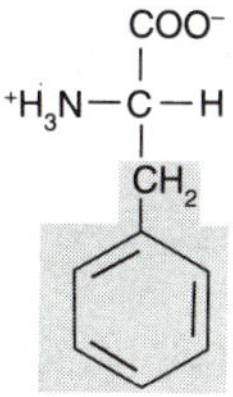

Phenylalanine (Phe, F) NP
Tyrosine (Tyr, Y) P UC — phenol
Tryptophan (Trp, W) NP — indole
Cysteine (Cys, C) P UC
Methionine (Met, M)

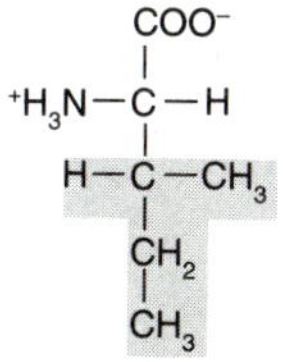

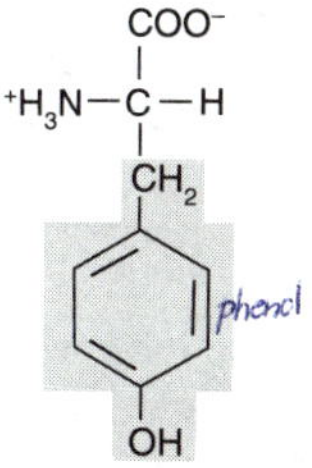

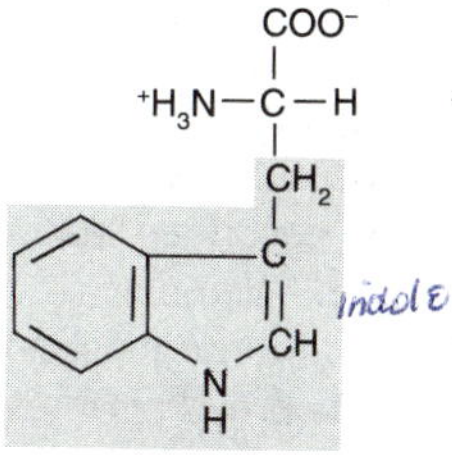

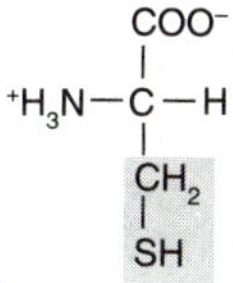

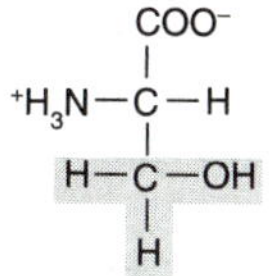

⎵ chymotrypsin ⎵

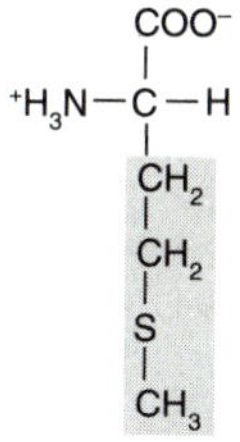

Serine (Ser, S) P UC
Threonine (Thr, T) P UC
Proline (Pro, P) NP
Aspartate (Asp, D) P C ⊖
Glutamate (Glu, E) P C ⊖

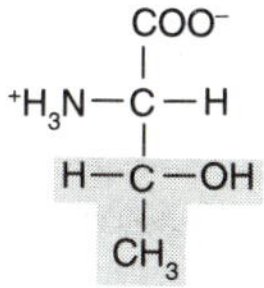

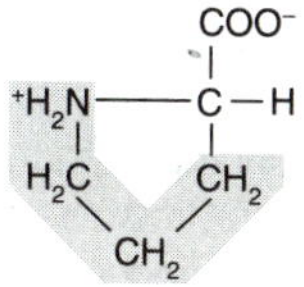

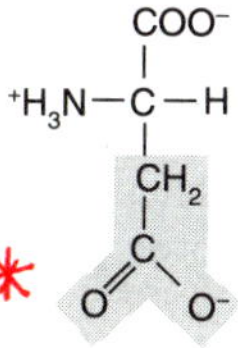

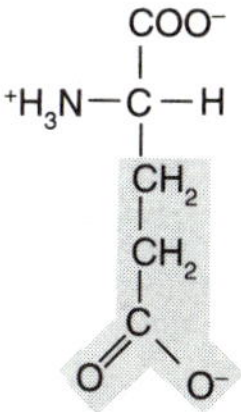

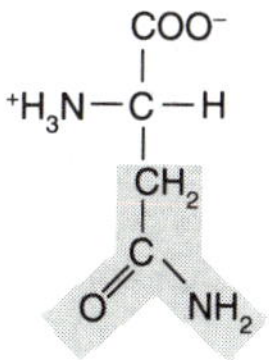

Asparagine (Asn, N) P UC
Glutamine (Gln, Q) P UC
Lysine (Lys, K) P C ⊕
Arginine (Arg, R) P C ⊕
Histidine (His, H) P C — imidazole

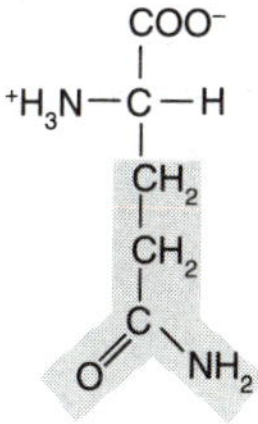

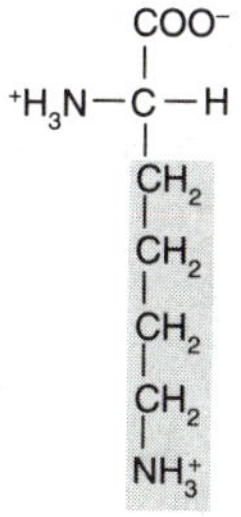

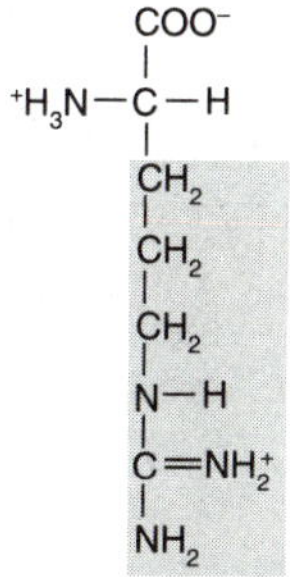

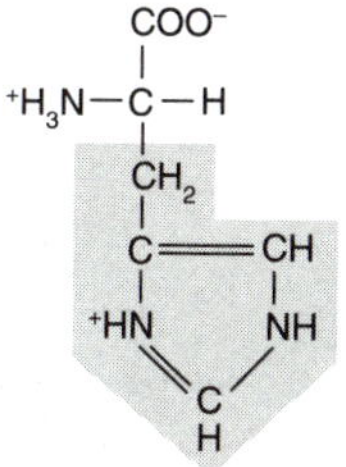

⎵ trypsin ⎵

a. β-alanine, a β-amino acid, is a building block of pantothenic acid, which is one of the structural units of coenzyme A (see Chapter 3, Bioenergetics)
b. γ-aminobutyric acid (GABA), a γ-amino acid, is involved in the transmission of nerve impulses

3. Because four different substituents are bonded to the α carbon of an amino acid, the α carbon is the chiral center and the amino acids (except glycine) are optically active
   a. Amino acids can be either D- or L-stereoisomers, depending on the orientation of substituents around the α carbon
   b. All amino acids found in proteins are L-amino acids; D-amino acids are found in certain bacterial cell walls and in some peptide antibiotics

4. The structure and function of an amino acid is determined to a large extent by the characteristics of its R group

5. One simple way to remember amino acids is to classify them as having one of three different types of side chains (R groups): amino acids with *nonpolar side chains,* amino acids with *charged polar side chains,* and amino acids with *uncharged polar side chains*
   a. The nine amino acids described below have a *nonpolar side chain* and various structures and sizes
      (1) *Glycine* (Gly, G), the simplest amino acid, has two identical substituents (two hydrogen atoms) bonded to the α carbon; because the α carbon is achiral, glycine is the only amino acid that is not optically active
      (2) *Alanine* (Ala, A) has a methyl group; *valine* (Val, V) has an isopropyl group; *leucine* (Leu, L) and *isoleucine* (Ile, I) contain **isomers** of butane
      (3) *Phenylalanine* (Phe, F) is so named because its R group can be described as a phenyl group attached to the methyl group of alanine
      (4) *Proline* (Pro, P) is unique in that its α-amino group, as well as its α carbon, are incorporated into a single cyclic structure; proline is more correctly classified as an *imino* acid rather than an amino acid, because its α-NH$_2$ group contains a secondary nitrogen (one that is bonded to two other alkyl (R) groups)
      (5) The R group of *tryptophan* (Trp, W) contains indole (a five-member, nitrogen-containing ring fused to a benzene ring)
      (6) *Methionine* (Met, M) contains sulfur in a thioether (R−S−R) linkage
   b. The five amino acids described below have *charged polar side chains*
      (1) Aspartate and glutamate are the only two common amino acids that bear a negative charge on the R group at neutral pH
         (a) *Aspartate* (Asp, D) contains a carboxyl group (−COOH) separated from the α carbon by a methylene (−CH$_2$−) group
         (b) *Glutamate* (Glu, E) contains a carboxyl group (−COOH) separated from the α carbon by two methylene groups
      (2) Lysine, arginine, and histidine are the only three common amino acids that bear a positive charge on the R group at neutral pH
         (a) *Lysine* (Lys, K) contains a butylamine side chain

(b) *Arginine* (Arg, R) contains a guanidinium group separated from the α carbon by three methylene groups

$$\text{NH}_2-\underset{\underset{\text{NH}_2}{\|}}{\text{C}}-\text{NH}-$$
(with $^+\text{NH}_2$ on the double-bonded nitrogen)

(c) *histidine* (His, H) contains a five-member heterocyclic ring, known as an *imidazole* group

c. The seven amino acids described below have *uncharged polar side chains*
  (1) *Serine* (Ser, S) contains a hydroxymethyl ($-CH_2OH$) group attached to the α carbon; *threonine* (Thr, T) contains ethanol attached to the α carbon; *tyrosine* (Tyr, Y) contains phenol (a benzene ring bearing a hydroxyl group); *cysteine* (Cys, C) contains a sulfhydryl ($-SH$) group
  (2) *Asparagine* (Asn, N) and *glutamine* (Gln, Q) are structurally similar to aspartate and glutamate, respectively, except that an amide group replaces the carboxyl group in the side chain of each; the designations Asx and Glx represent the sum of aspartate and asparagine, or glutamate and glutamine, in a protein, respectively

C. Chemical properties of amino acids

1. The chemical properties of the R group determine the chemical properties of each amino acid
2. The number of ionizable groups in an amino acid determine the extent to which it will be ionized at a given pH in the body
3. Amino acids may be categorized according to the functional groups and chemical properties of their side chains; some amino acids groups fit more than one classification
   a. The amino acids with aliphatic (non-ring) side chains are glycine, alanine, valine, leucine, and isoleucine
      (1) Glycine and alanine have the smallest aliphatic side chains; they play an important structural role in proteins by fitting into spaces too small for a larger amino acid
      (2) Valine, leucine, and isoleucine have longer aliphatic side chains that are hydrophobic; they tend to congregate in structures that enable them to avoid water
   b. The aromatic (benzene-related) amino acids, phenylalanine, tyrosine, and tryptophan, all have the aromatic group attached to the α carbon via a methylene ($-CH_2-$) group
      (1) Phenylalanine and tryptophan are hydrophobic and tend to form structures that enable them to avoid water
      (2) The hydroxyl group on tyrosine makes it more hydrophilic and a more chemically reactive acid than phenylalanine
      (3) Although histidine and proline contain ring structures, they are usually classified differently
   c. Cysteine and methionine, the two sulfur-containing amino acids, are hydrophobic

(1) Two cysteine residues can bond via their thiol (−SH) groups to form the derived amino acid cystine;

$$\begin{array}{cc} CH_2-S-S-CH_2 \\ | & | \\ HC-NH_2 & HC-NH_2 \\ | & | \\ CO_2H & CO_2H \end{array}$$

the thiol group of cysteine is chemically reactive and easily forms a disulfide bond (−S−S−) through oxidation

(2) The formation of disulfide bonds is important in protein structures; disulfide bonds commonly join two different peptide chains together or link two different cysteine residues within the same peptide

d. Serine and threonine have a side chain hydroxyl group that makes them both hydrophilic and chemically reactive

e. In proline, the incorporation of nitrogen from the α-amine group into the side chain constrains the rotation of the molecule around the C−N bond; this bond constraint affects the structure of a protein causing it to bend at the position of a proline residue

f. The acidic amino acids aspartate and glutamate are dicarboxylic, monobasic amino acids because they have two carboxyl groups (one bound to the α carbon and one in the side chain) but only one amine group (bound to the α carbon); the carboxyl groups of both amino acids are negatively charged at physiologic pH (7.4)

g. Amino acids with basic side chains are lysine, arginine, and histidine

(1) The side chain amine group makes these amino acids extremely polar and hydrophilic; at physiologic pH, both lysine and arginine have positively charged amine groups

(2) At physiologic pH, the imidazole ring in histidine exists primarily in the uncharged form because the imidazole ring has a pK'_a of approximately 6

h. Asparagine and glutamine, the two amino acids that contain an amide group in the side chain, are not charged at physiologic pH

D. Ionization of amino acids

1. All free amino acids contain at least two ionizable groups, the α-carboxyl group and the α-amine group; the charge on amino acid R groups within a protein affects the three-dimensional structure and reactivity of the protein

2. Except for the first and last amino acids in a protein, all of the α-amine and α-carboxyl groups within a protein participate in peptide bonds and thus do not contribute to the protein's overall charge

3. The overall charge of a protein is determined by the charge on the amino acid side chains (the charge on the R groups) and the charge on the protein's N-terminal and C-terminal residues (see *A Small Peptide of Six Amino Acids*, page 22)

a. The pK'_a of the α COOH is 2.35; at physiologic pH, the proton is removed and the carboxyl group is in the ionized form with a −1 charge

b. The pK'_a of the α NH_2 is 9.69; at physiologic pH, the NH_2 is protonated to the ionized form ($-NH_3^+$) with a +1 charge
4. All amino acids contain both an acidic group (the α-carboxyl group) and a basic group (the α-amine group) in the same molecule
   a. An amino acid without an ionizable side chain will have one +1 charge (from the α-amine group) and one −1 charge (from the α-carboxyl group) at physiologic pH; its net charge is zero
   b. A charged compound with a net charge of zero is called a **zwitterion**
5. The pH at which a compound exists as a zwitterion is its **isoelectric point** (pI)
6. The pH of the environment determines the degree of ionization of an amino acid; this is illustrated in the titration of the two ionizable groups of alanine (see *Titration Curve for Alanine*, page 28)
   a. At pH < 2.35, the two ionizable groups (the α COOH and the α NH_2) are completely protonated; the carboxyl group is un-ionized and exists as −COOH; the amine group has a +1 charge and exists as $-NH_3^+$
   b. At pH = 2.35, which is the pK'_a of the carboxyl group, equal amounts of the ionized ($-COO^-$) and un-ionized (−COOH) carboxyl group are present
   c. As the pH increases beyond 2.35, the carboxyl group on most of the molecules remains unprotonated and thus is negatively charged; the amine group is still protonated because the pH has not reached the amine pK'_a of 9.69, and hence the amine group on most molecules is still charged
   d. At the pH midway between the two pK'_a values, the negative charge from the carboxyl group equals the positive charge from the amine group; this pH ([2.35 + 9.69] ÷ 2 = 6.02) is the isoelectric point of alanine, the pH at which alanine exists as a zwitterion
   e. At pH = 9.69, which is the pK'_a of the amine group, equal amounts of the un-ionized ($-NH_2$) and ionized ($-NH_3^+$) amine groups are present; the carboxyl group, well beyond its p$K'a$, remains ionized ($-COO^-$)
   f. At pH > 9.69, none of the ionizable groups of alanine are protonated; the carboxyl group has a negative charge ($-COO^-$) and the amine group is un-ionized ($-NH_2$)
7. An ionizable group in the side chain of an amino acid has its own pK'_a; in these amino acids, the side chain pK'_a also affects the isoelectric point

II. Levels of Protein Structure

A. General information
1. All proteins have at least three levels of structure: primary, secondary, and tertiary
2. Proteins composed of more than one polypeptide chain also have a quaternary structure
3. *Monomers* are proteins composed of only one polypeptide chain; proteins composed of two or more polypeptide chains are called *oligomers* and are classified by the number of polypeptide chains; dimers have two chains; trimers, three chains; tetramers, four chains, and so forth

Titration Curve for Alanine

The titration of alanine shows that the isoelectric point is equal to one-half of the sum of the pK'_a values of the amino acid (which represents the two ionizable groups of the amino acid). $pI = \frac{1}{2} \times (9.69 + 2.35) = 6.02$. The isoelectric point is the pH at which the amino acid exists as a zwitterion.

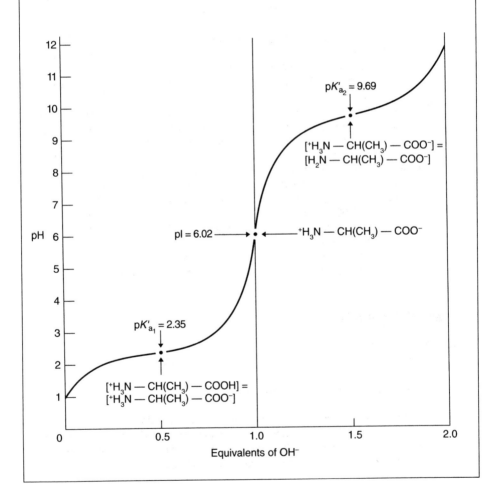

Formation of a Peptide Bond

A peptide bond is the amide linkage that results from the condensation reaction between the α-carboxyl group of one amino acid and the α-amine group of another amino acid. In the reaction shown below, the participating α-carboxyl and α-amine groups are enclosed.

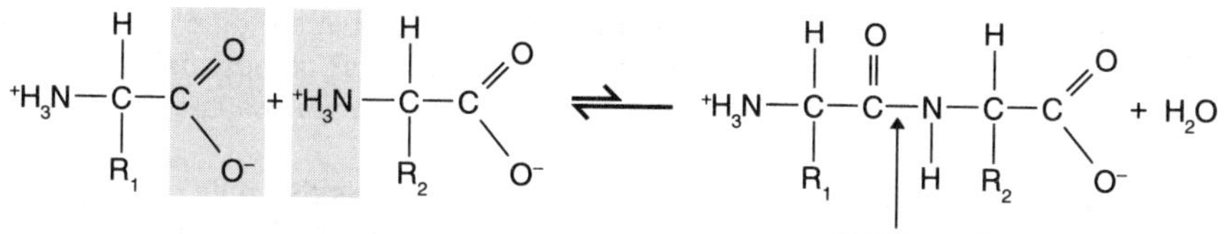

$$^+H_3N-\underset{R_1}{\underset{|}{C}}-\underset{O^-}{C\overset{O}{\diagup\!\!\!\diagdown}} + {}^+H_3N-\underset{R_2}{\underset{|}{C}}-\underset{O^-}{C\overset{O}{\diagup\!\!\!\diagdown}} \rightleftharpoons {}^+H_3N-\underset{R_1}{\underset{|}{C}}-\underset{}{\overset{O}{\underset{||}{C}}}-\underset{H}{\underset{|}{N}}-\underset{R_2}{\underset{|}{C}}-\underset{O^-}{C\overset{O}{\diagup\!\!\!\diagdown}} + H_2O$$

Peptide bond

4. The primary, secondary, tertiary, and quarternary structure of a protein controls its three-dimensional shape, which partly determines its activity

B. Primary structure of protein
1. Amino acids are linked by peptide bonds (see *Formation of a Peptide Bond*)
    a. A peptide bond forms as result of a condensation reaction between the carbonyl carbon from the α COOH of one amino acid and the amine nitrogen from the α NH_2 of the next amino acid; the reaction eliminates a molecule of water
    b. The peptide bond is rigid and planar; it has a bond length intermediate between that of a single bond and a double bond
    c. The partial double-bond character of the peptide bond restricts its ability to rotate, making it the "backbone" of the protein
    d. Other bonds in amino acids are single bonds and therefore have more freedom of rotation than peptide bonds
2. The primary structure of a protein is the sequence and number of amino acids linked by peptide bonds
    a. The convention for writing the primary structure of an amino acid is to name its sequence of amino acids from left to right
    b. The first amino acid (the N-terminal amino acid) has a free α-amine group; the carboxyl group of this first amino acid participates in formation of the first peptide bond
    c. Subsequent amino acids are written to the right of the N-terminal amino acid; all of these residues are linked by peptide bonds at their α-amine and α-carboxyl groups
    d. The last amino acid (the C-terminal amino acid) has a free α-carboxyl group (see *A Small Peptide of Six Amino Acids*, page 22)

C. Secondary structure of protein
1. Intramolecular and intermolecular hydrogen bonding in the primary structure give the protein its secondary structure
2. The α helix, the β pleated sheet, and the collagen triple helix are three different types of secondary structure
3. The α helix occurs in many different fibrous and globular proteins; it is formed by *intramolecular* hydrogen bonds between the carboxyl group oxygen of one amino acid and the amine hydrogen of the fourth amino acid in front of it

a. Every carboxyl oxygen and amine hydrogen participates in the intramolecular bonding; the combined forces of all these hydrogen bonds give the protein an overall rodlike structure
   b. Formation of the α helix occurs spontaneously in a pattern of 3.6 amino acids per turn of the α helix
   c. Amino acids that are three to four residues apart in the primary structure are spatially close to each other in the α helix; R groups that are adjacent to each other in the primary structure are spatially distant in the α helix
   d. The extent of α-helical structure in a protein depends on its constituent amino acids; some amino acids (Leu, Ala, Glu, Met) tend to favor α helix formation, whereas others (Val, Ser, Asp, Asn) tend to destabilize it
4. The β pleated sheet is the secondary structure of silk, a fibrous protein produced by spiders and insects; it is formed by *intermolecular* hydrogen bonds between a carboxyl group oxygen in one amino acid and an amine group hydrogen in an adjacent polypeptide
   a. A parallel β pleated sheet is formed when adjacent polypeptides are oriented in the same direction (from N-terminal to C-terminal or vice versa)
   b. An antiparallel β pleated sheet is formed when adjacent polypeptides are oriented in opposite directions (one chain oriented from the N- to C-terminal and the adjacent chain oriented from the C- to N-terminal)
   c. Proteins that have β pleated sheets assume an extended, rather than a rod-shaped or coiled, structure
5. Collagen, the most abundant structural protein found in vertebrates, is a major component of connective tissues, such as bone, teeth, cartilage, and tendon; the secondary structure of collagen is a triple helix composed of three entwined polypeptide chains and bound by intermolecular hydrogen bonds
   a. The collagen triple helix has a high concentration of proline and glycine; often every third residue in each of the three chains is a glycine molecule
   b. Two derived amino acids, hydroxyproline and hydroxylysine, occur frequently; the sequence — Gly — Pro — hydroxyproline — occurs frequently
   c. There are 3.3 amino acids per turn of the collagen triple helix, with approximately every third residue projecting into the helix interior
   d. The high concentration of glycine (the smallest amino acid) prevents crowding in the interior of the helix

D. Tertiary structure of protein
1. A protein's tertiary structure results from *interactions* between amino acid side chains of each of the residues; these interactions between R groups cause a polypeptide to assume its normal three-dimensional configuration
2. Many types of noncovalent interactions are involved, such as van der Waals forces, ionic bonding, hydrophobic bonding, and hydrogen bonding
   a. Residues with hydrophilic side groups usually are located on the exterior of the molecule, oriented to interact with the aqueous environment
   b. Residues with hydrophobic side groups usually are located on the interior of the molecule, oriented so the hydrophobic groups can interact with each other and exclude water

3. One type of covalent bond that influences tertiary structure is the disulfide bond of the derived amino acid cystine (see section I. C. 3. c. (1).)

E. Quaternary structure of protein
1. Most proteins are composed of more than one polypeptide chain; these polypeptide subunits associate with each other in a defined geometric arrangement called the quaternary structure
2. Formation of the quaternary structure involves the same noncovalent interactions as in the formation of a tertiary structure, but the interactions occur between polypeptide chains (interchain) rather than within a polypeptide chain (intrachain)
3. Oligomeric proteins composed of identical subunits are called homogeneous; those composed of different subunits are called heterogeneous

III. Techniques for Studying Proteins

A. General information
1. The amino acid *composition* of a protein can be discerned by cleaving all the peptide bonds, then separating and identifying the constituent amino acids
2. The amino acid *sequence* of a protein can be discerned by using various methods to cleave only selected peptide bonds, then assimilating all the information to deduce the amino acid sequence; this process is similar to putting together the pieces of a puzzle
3. Proteins are also identified and characterized by spectroscopy, ultracentrifugation, chromatography, and electrophoresis
4. More sophisticated aspects of protein characterization are accomplished by X-ray diffraction, circular dichroism spectroscopy, nuclear magnetic resonance spectroscopy, and mass spectroscopy

B. Determining amino acid composition and sequence
1. In the first stage of determining the amino acid composition of a protein, all the peptide bonds are hydrolyzed by an acid or a base
   a. In acid hydrolysis, the protein solution is usually treated with 6 N hydrochloric acid at 110° C for 24 hours
   b. In base hydrolysis, the protein solution is usually treated with 2 to 4 N sodium hydroxide at 100° C for 4 to 8 hours
   c. Under acid hydrolysis, aspartate cannot be distinguished from asparagine and glutamate cannot be distinguished from glutamine; the totals for both types of molecules are identified as Asx and Glx, respectively
   d. Trytophan is acid-labile and requires base hydrolysis for identification
2. In the second stage of determining the amino acid composition of a protein, the hydrolyzed amino acids are separated, identified, and quantitated
   a. Amino acids can be separated by thin layer chromatography or liquid chromatography, which is the separation of molecules on a solid support according to their physical or chemical properties
   b. Amino acids can be identified and quantitated by reacting them with reagents that produce a colored or fluorescent product, the intensity of which is proportional to the concentration of amino acid present

(1) Ninhydrin, a reagent that turns blue when reacted with amino acids and yellow when reacted with imino acids (proline), is sensitive to microgram (10^{-6} g) quantities of amino acids

(2) Fluorescamine, a reagent that reacts with the α-amine group to produce a fluorescent derivative, is more sensitive than ninhydrin; fluorescamine can detect nanogram (10^{-9} g) quantities of amino acids

3. After determining the amino acid composition of the protein, the determination of the structure includes identifying the N-terminal residue and C-terminal residue

   a. The N-terminal amino acid is identified by the reaction of its free amine group in the intact protein
      (1) The reagent fluorodinitrobenzene (FDNB) reacts with the free amine group of a protein to form a covalent dinitrophenyl derivative
      (2) The protein is subjected to acid hydrolysis and chromatographic separation as previously described
      (3) The dinitrophenyl derivative has different chromatographic properties than the other amino acids and can thus be identified
   b. The free amine group in the side chain of lysine also reacts with FDNB and must be considered when interpreting results
   c. An N-terminal amino acid that is modified on the amine group will not have a free amine group to react with FDNB
   d. Other reagents, such as dabsyl chloride and dansyl chloride, frequently replace FDNB; they permit detection of smaller quantities of the amino acid
   e. The *Edman degradation* is a method of identifying an entire sequence of amino acids in a protein by successively releasing individual N-terminal residues that can be separated from the intact protein and identified
      (1) The reagent phenyl isothiocyanate (PITC, also called Edman's reagent) reacts with the N-terminal amino acid to form a thiazolinone derivative, which is then extracted with organic solvent and converted to a more stable molecule called a phenylthiohydantoin (PTH) amino acid
      (2) The technique used in the Edman degradation separates the N-terminal residue only and *leaves the rest of the polypeptide intact*
      (3) The separated PTH amino acid is then identified using standard identification techniques, such as chromatography and electrophoresis
      (4) The next amino acid is the new N-terminal amino acid and is sequentially labeled and removed without altering the remaining protein
      (5) The cycles are repeated, with a new N-terminal amino acid generated during each cycle, until all amino acids have been separated and identified in sequence
   f. The identification of the C-terminal residue in a protein is accomplished by treating the intact protein with a carboxypeptidase, a class of enzymes that selectively catalyze the removal of the C-terminal amino acid from the peptide

(1) Other enzymes that cleave specific peptide bonds on their carboxyl side include trypsin, which cleaves at residues that contain positively charged R groups (Lys and Arg), and chymotrypsin, which cleaves at residues with aromatic R groups (Phe, Tyr, and Trp) and bulky aliphatic R groups (Ile and Val)
(2) Cyanogen bromide is a chemical that cleaves at the carboxyl side of the peptide bond of methionine

C. Dissociation of protein subunits
1. Proteins with multiple subunits must be broken down to individual subunits to characterize their amino acid composition and sequence
2. *Denaturation* is the process of treating a protein with heat or chemical reagents to break it down into its constituent subunits
    a. Denaturation causes loss of a protein's secondary, tertiary, and quaternary structure, but its primary structure may be maintained
    b. *In vivo,* changes in pH, ionic strength, and temperature denature proteins; *in vitro,* detergents, urea, organic solvents, acids, or bases denature proteins
    c. Denatured proteins lose their normal biochemical activity and are nonfunctional
    d. Disulfide bonds in proteins are disrupted when reducing agents, such as β-mercaptoethanol or dithiothreitol, are added

D. Identification, separation, and characterization of proteins
1. The presence of a protein and its integrity can be detected by *spectroscopy,* a technique for identifying and quantitating unknown molecules by virtue of their interaction with light
    a. Absorption spectroscopy identifies peptide bonds and certain functional groups by measuring the absorbance of different wavelengths of ultraviolet (UV) or visible light
        (1) Peptide bonds absorb UV light in the range of 180 to 230 nanometers (nm)
        (2) The aromatic rings in phenylalanine, tryptophan, and tyrosine absorb UV light in the range of 260 to 300 nm
        (3) Denaturation changes in secondary and tertiary structure appear as a difference in light absorbance between 180 and 230 nm
    b. Fluorescence spectroscopy, a more sensitive measurement than UV or visible spectroscopy, detects the presence and quantity of a protein
        (1) Fluorescence occurs when a molecule in an excited state, produced from absorption of short wavelength (high energy) light, re-emits some of its excess energy in the form of longer wavelength (lower energy) light; the excess energy not emitted as light dissipates as heat
        (2) Only proteins containing amino acids with aromatic side chains (Phe, Tyr, Trp) will naturally fluoresce
        (3) A protein that does not naturally fluoresce can be made to fluoresce by covalently bonding a fluorescent group to the protein
2. *Ultracentrifugation* separates proteins according to size

a. The sedimentation coefficient of a protein, subjected to a centrifugal (outward) force, depends on the protein's mass, density, and shape and on the density of the suspending solution
b. The Svedberg (S) unit, a measure of the sedimentation coefficient of a particle, is equal to 10^{-13} second
c. The magnitude of the sedimentation coefficient gives a relative value that can characterize the molecular weight of a protein

3. *Gel filtration chromatography* (also called size exclusion chromatography) separates proteins according to size
   a. Inert porous beads are packed into a column; the pore size is selected to correspond to the molecular weight of molecules small enough to enter the pores
   b. A buffered protein solution is then applied to the top of the packed column
   c. High molecular weight proteins are too large to enter the pores in the beads and thus go directly through the column
   d. Low molecular weight proteins enter the pores and thus have a longer path to traverse on their way through the column; some midsize proteins can enter the pores while others cannot; they travel through the column at rates between high and low molecular weight proteins
   e. The elution time of each protein is related to its molecular size; large proteins elute first, midsized proteins elute next, and small proteins elute last

4. *Ion exchange chromatography* separates proteins based on their charge
   a. Inert beads coated with either negatively or positively charged groups are packed into a column, and a buffered protein solution is then applied to the top of the packed column
   b. Proteins with a net positive charge will bind, via ionic bonds, to beads coated with negatively charged groups (**cation** exchange); conversely, proteins with a net negative charge will bind to beads coated with positively charged groups (**anion** exchange)
   c. The addition of buffer solution containing a higher concentration of the same type of ions as that in the original protein solution displaces, or exchanges, the protein's ionic bonding to the charged groups on the beads; this allows the researcher to control when the protein of interest elutes from the column

5. *Affinity chromatography* separates proteins that bind specifically with certain chemical groups
   a. Inert beads coated with specific chemical groups are packed into a column; a buffered protein solution is then applied to the top of the packed column
   b. A protein with an affinity for the specific chemical group will bind to that group and consequently will be retained on the column; proteins that lack affinity will not bind and will flow directly through the column
   c. Addition of a higher concentration of a soluble form of the specific chemical group displaces the protein from the binding sites on the beads; the protein then elutes separately

6. *Electrophoresis*, the separation of charged molecules by their movement on a solid support while exposed to an electric field, also aids in the separation of different proteins

a. Electrophoresis separates proteins on the basis of their net charge or on the basis of their mass if they can be made to have the same charge
b. The anionic detergent sodium dodecyl sulfate (SDS) is added to a protein solution; SDS binds strongly to most proteins, causing denaturation and imparting a large net negative charge to each protein
c. The large negative charge assumed by each protein masks its intrinsic charge; each protein in the SDS solution tends to have the same charge-to-mass ratio and a similar shape
d. The SDS-treated proteins are placed on the solid support, along with several marker proteins of known molecular weight, and are exposed to an electric field; the proteins vary in the distance they move on the support in direct relation to their molecular weight
7. *Isoelectric focusing* separates proteins on the basis of their isoelectric points
   a. The support medium is a gel in which a pH gradient has been established
   b. The protein mixture is applied to the gel; proteins migrate through the gel until they reach the pH at which they have no net charge (their pI); with no net charge, proteins cease to migrate
   c. Because each protein has its own pI, separation of a mixture of proteins can be achieved

E. **Determination of higher levels of protein structure**
   1. The three-dimensional structure of a protein (its secondary, tertiary, and quaternary structure) may be determined by X-ray diffraction
   2. The amount and type of secondary structure in a protein is determined by circular dichroism (CD) spectroscopy
   3. Nuclear magnetic resonance spectroscopy (NMR) provides structural and functional information about the environment of a particular atom in a protein
   4. Mass spectroscopy (MS) can determine the structure of very small quantities of protein

Study Activities

1. Draw the structures of each of the 20 common amino acids; identify the composition and chemical reactivity of the side chains.
2. List the forces involved in the formation of primary, secondary, tertiary, and quaternary structures of proteins.
3. Describe the steps by which the primary structure of a protein is determined.
4. Define the basic principle of each of the following techniques: spectroscopy, ultracentrifugation, electrophoresis, and gel filtration chromatography.
5. Describe how the Edman degradation technique can determine the entire primary structure of a protein.
6. Define the terms listed in bold-faced type throughout this chapter.

3

Bioenergetics

Objectives

After studying this chapter, the reader should be able to:
- Name the high energy compounds and identify the sites at which they are produced and consumed in glycolysis, the tricarboxylic acid cycle, and the respiratory chain.
- Trace the path for the complete oxidation of one molecule of glucose to carbon dioxide and water, identifying all intermediate products, enzymes, cofactors, and major regulatory sites
- Discuss the chemical reactions of glycolysis and the tricarboxylic acid cycle in terms of their substrates, products, enzymes, cofactors, and regulatory mechanisms.
- Explain the process by which oxidative phosphorylation produces ATP molecules in the respiratory chain.
- Discuss inhibitors of the respiratory chain and uncouplers of oxidative phosphorylation.

I. Energy Requirements for Metabolism

A. General information
1. Metabolism encompasses the chemical processes used to synthesize complex molecules from basic precursor molecules (anabolism) and to break down complex molecules into less complex components (catabolism)
2. Anabolism consumes energy; catabolism generates energy
3. The change in *Gibbs free energy* (ΔG) is a criterion used to predict whether or not a given chemical reaction will be spontaneous (proceed without outside energy input)
   a. Although it is impossible to measure the absolute value of free energy of a given molecule, it is possible to measure the change in total free energy, written ΔG, of a system (a collection of molecules at a given concentration, temperature, pressure, and pH) as a result of a particular reaction
   b. In biochemistry, the change in Gibbs free energy under standard conditions (when all components are at a concentration of 1 molar, 25° C, 1 atm pressure, and a pH of 7) is indicated by $\Delta G°'$
   c. The commonly used units for $G°'$ are kilocalories per mole of reactant (kcal/mol) or kilojoules per mole of reactant (kJ/mol); 1 kcal = 4.184 kJ

4. An ***exergonic*** reaction is any reaction that releases Gibbs free energy, has a negative value for ΔG, and occurs spontaneously; an ***endergonic*** reaction is one that requires an input of Gibbs free energy, has a positive value for ΔG, and does not occur spontaneously
5. In a series of metabolic reactions, the overall Gibbs free energy change equals the sum of the Gibbs free energy changes for each individual reaction

$$(1)\ A + B \leftrightarrow C + D\quad \Delta G_1$$
$$(2)\ D + E \leftrightarrow F + G\quad \Delta G_2$$
$$(1) + (2)\ A + B + E \leftrightarrow C + F + G\quad \Delta G_3$$
$$\Delta G_3 = \Delta G_1 + \Delta G_2$$

   a. If the overall free energy change for a series of coupled reactions is negative (if $\Delta G_3 < 0$), the overall reaction is exergonic and occurs spontaneously in the forward direction
   b. If $\Delta G_3 > 0$, the overall reaction is endergonic and does not operate spontaneously in the forward direction
   c. If $\Delta G_3 = 0$, the reaction is at equilibrium and the rate of the forward reaction equals the rate of the reverse reaction
6. In a series of metabolic reactions, an endergonic reaction can occur if a sufficiently exergonic reaction is coupled to it; as long as the overall Gibbs free energy change of a reaction series is negative, the reactions in the series can occur spontaneously

B. Sources of free energy
1. The hydrolysis of adenosine triphosphate (ATP) provides the main source of free energy in biochemical reactions
2. The triphosphate group of ATP contains two high-energy phosphoanhydride bonds that, when hydrolyzed, each release approximately 7.3 kcal/mol of free energy; $\Delta G^{\circ\prime}$ for ATP hydrolysis to adenosine diphosphate (ADP) = -7.3 kcal/mol
3. ATP is actually a free energy transmitter rather than a free energy reservoir; the amount of ATP in a typical cell is usually only enough to meet free energy needs for a minute or two; to meet ongoing needs for free energy, cells continually regenerate new ATP
4. Cells replenish ATP through three biochemical pathways: ***glycolysis,*** the ***tricarboxylic acid (TCA) cycle,*** and ***oxidative phosphorylation***
   a. In glycolysis (see section II. of this chapter), cells oxidize glucose to form pyruvate or (under anaerobic conditions) lactate; in the overall conversion of glucose to either pyruvate or lactate, each molecule of glucose provides enough energy to form two molecules of ATP
   b. In the TCA cycle (also called the Krebs cycle or the citric acid cycle), cells first convert pyruvate to acetyl coenzyme A (acetyl CoA) and then oxidize it completely to CO_2; the oxidation of two pyruvate molecules provides enough energy to form two molecules of the high-energy compound GTP, which can be converted to ATP
   c. Oxidative phosphorylation is the primary source of ATP for living cells
      (1) During oxidative phosphorylation, cells transfer the electrons resulting from the complete oxidation of glucose to special

acceptor molecules called cofactors, nicotinamide adenine dinucleotide (NAD^+) and flavin adenine dinucleotide (FAD^+)
(2) Electrons then move from the cofactor molecules through a series of enzyme complexes called the electron transport chain or respiratory chain and ultimately to oxygen to form H_2O; this movement of electrons, actually a series of oxidation-reduction reactions, yields enough energy to produce 32 or 34 molecules of ATP for each molecule of oxidized glucose

d. Each complete oxidation of one glucose molecule to CO_2 and H_2O yields either 36 or 38 molecules of ATP (from the steps described in a., b., and c., above)

e. The interconversion of adenosine monophosphate (AMP), ADP, and ATP is catalyzed by adenylate **kinase:** AMP + ATP $\rightleftarrows$ ADP + ADP; this reaction can produce ATP when it is needed for thermodynamically unfavorable reactions

5. Other nucleoside triphosphates (NTPs) such as guanosine-, uridine-, and cytidine triphosphate (GTP, UTP, and CTP, respectively) also yield free energy through the hydrolysis of their phosphoanhydride bonds

    a. ATP can transfer one or both of its two high-energy phosphate groups to other nucleoside monophosphates (NMPs) or nucleoside diphosphates (NDPs)

    b. A nucleoside monophosphate **kinase** catalyzes the reaction NMP + ATP $\rightleftarrows$ NDP + ADP; a nucleoside diphosphate kinase catalyzes the reaction NDP + ATP $\rightleftarrows$ NTP + ADP

6. Another high-energy molecule, acetyl CoA transfers acetyl groups in a way that is analogous to the transfer of phosphoryl groups by ATP

    a. The terminal sulfhydryl group, the active site in acetyl CoA, links the acetyl group to CoA by a high energy thioester bond ($\sim S$)

    b. CoA is a derivative of pantothenate, a B-complex vitamin

    c. $\Delta G^{\circ\prime}$ for hydrolysis of acetyl CoA to acetate and CoA is -7.5 kcal/mol

C. Release of free energy by oxidation of foodstuffs

1. The human body obtains the energy needed for all its metabolic processes through the oxidation of carbohydrates, fats, and proteins in food

    a. Initially, complex carbohydrates are catabolized to simple sugars, fats are catabolized to glycerol and fatty acids, and proteins are catabolized to amino acids

    b. These simple sugars, fatty acids, and amino acids are further catabolized to a few common metabolic intermediates, such as acetyl units, pyruvate, or α-ketoglutarate; the acetyl units are carried into the TCA cycle as acetyl CoA

2. Acetyl groups enter the TCA cycle as acetyl CoA; the TCA cycle generates four pairs of electrons that are carried to the respiratory chain where they are oxidized to carbon dioxide (CO_2) and water; it also produces ATP simultaneously

3. Electrons are transferred in the electron transport chain from NADH and $FADH_2$ to special electron-carrier molecules; at the end of the chain, electrons are transferred to molecular oxygen to form H_2O; as electrons move through the chain, ATP is produced from ADP and inorganic phosphate (P_i)

4. **Oxidative phosphorylation** is the synthesis of ATP from ADP and P_i that occurs during the transport of electrons from NADH and $FADH_2$ to oxygen

D. **Regulation of metabolism**
   1. Prokaryotes regulate metabolism at the level of deoxyribonucleic acid (DNA) transcription (see Chapter 7, Transmission of Genetic Information); synthesis of the metabolic enzymes takes place in response to the cell's immediate needs; for example, prokaryotes synthesize the enzymes needed to metabolize lactose only if lactose is available and glucose is scarce (see Chapter 8, Gene Regulation and Analysis)
   2. Eukaryotes use three primary routes for regulation of metabolism: enzyme control, differentiation of anabolic and catabolic pathways, and physical separation of metabolic pathways
      a. Eukaryotes control the catalytic activity of enzymes by regulating the rate of enzyme synthesis and degradation through the use of allosteric inhibitors and activators (see Chapter 4, Protein Function and Metabolism), and through **covalent modification** of the enzyme
      b. Although there are many exceptions to the rule, eukaryotes have different chemical pathways for anabolic and catabolic reactions, especially at key regulatory steps; pathway differentiation permits separate but cooperative regulation of catabolism and anabolism and prevents energy waste caused by both paths operating simultaneously
      c. Anabolic and catabolic reactions generally are segregated in different parts of the cell; for example, catabolism of fatty acids takes place in the mitochondria while anabolism of fatty acids takes place in the cytosol
   3. The **energy charge** of a cell is the ratio of cellular ATP relative to AMP and ADP; cells maintain the energy charge within a narrow range; the balance of ATP versus AMP and ADP dictates whether ATP is required and must be synthesized or whether ATP is abundant (ready to be used)

II. Glycolysis

A. **General information**
   1. *Glycolysis*, a nearly universal pathway in prokaryotes and eukaryotes, is the series of enzyme-catalyzed reactions by which one molecule of glucose is converted to two molecules of pyruvate; two ATP molecules also are produced
      a. Because glycolysis requires no oxygen, glycolytic reactions are the same in the presence of oxygen (aerobic conditions) or in its absence (anaerobic conditions)
      b. The presence or absence of oxygen effects the fate of pyruvate, the final product of glycolysis
      c. Anaerobic conditions cause pyruvate to be converted to lactate in a reaction called *homolactic fermentation*
      d. In yeast and certain bacteria, anaerobic conditions cause *alcoholic fermentation* of pyruvate to form ethanol and CO_2 (the basis of wine production)
      e. Under aerobic conditions, pyruvate is converted to acetyl CoA and enters

the TCA cycle; electrons produced in the TCA cycle then enter the electron transport chain, where more ATP is produced
  f. The glycolytic pathway also produces intermediate products needed for the synthesis of other macromolecules, such as proteins, lipids, and other carbohydrates
 2. The complete oxidation of glucose through glycolysis, the TCA cycle, and oxidative phosphorylation produces energy that is stored in the form of ATP
    a. In glycolysis, the conversion of one molecule of glucose to two molecules of pyruvate produces two molecules of ATP
    b. The complete oxidation of two molecules of pyruvate to CO_2 through the TCA cycle produces two molecules of ATP
    c. The process of oxidative phosphorylation produces 32 or 34 molecules of ATP for every molecule of glucose that began glycolysis (depending on the shuttle used)
 3. In glycolysis, all the intermediate molecules between glucose (the starting compound) and pyruvate or lactate (the final products) are phosphorylated; the phosphate group is added as an ester or an anhydride; anhydride bonds are usually high-energy bonds

B. The glycolytic pathway
 1. The first series of reactions in glycolysis, called the *priming stage,* convert glucose to fructose-1,6-bisphosphate (for an illustration of all stages in glycolysis, see *Glycolytic Pathway,* pages 42 and 43)
    a. Glucose enters cells by a specific transport mechanism and is phosphorylated in *Step 1* to yield glucose 6-phosphate (glucose 6-P); this phosphorylation is the first irreversible reaction of glycolysis and one of three **rate-limiting steps** of glycolysis
    b. Phosphorylation converts glucose to an ionic molecule that is not able to penetrate the lipid bilayer membrane, thus trapping glucose 6-P inside the cell for subsequent reactions
    c. The phosphate group, provided by the hydrolysis of ATP to ADP, consumes the first molecule of ATP for the overall reaction
    d. *Hexokinase,* the enzyme catalyzing this reaction, resembles all kinases in that it requires a divalent metal ion, usually magnesium (Mg^{2+}), as a cofactor
    e. Hexokinase is inhibited by glucose 6-P, the product of the reaction it catalyzes
    f. Glucokinase, an **isoenzyme** of hexokinase, also catalyzes the conversion of glucose into glucose 6-P but has a higher K_m (lower affinity) for glucose; it provides glucose 6-P to be used in the synthesis of glycogen (see Chapter 5, Carbohydrate Structure, Function, and Metabolism)
    g. Once glucose 6-P is formed, it can proceed through glycolysis, be converted to glucose 1-P for glycogen synthesis, or be oxidized in the pentose phosphate pathway (see Chapter 5, Carbohydrate Structure, Function, and Metabolism)
    h. In *Step 2* of glycolysis, glucose 6-P is isomerized to fructose 6-P; *phosphoglucose isomerase* catalyzes this reversible reaction
    i. In *Step 3,* fructose 6-P is further phosphorylated to fructose 1,6-bisphosphate (fructose 1,6-bisP); this is the second irreversible reaction

of glycolysis and is catalyzed by *phosphofructokinase (PFK);* the phosphate group, provided by the hydrolysis of ATP to ADP, consumes the second molecule of ATP
   j. Step 3 is a unique step in glycolysis and its enzyme, PFK, the most important regulator of glycolysis and is highly controlled
      (1) PFK catalyzes the **commitment step** for glycolysis because its product, fructose 1,6-bisP, is found in no other metabolic pathway; once fructose 1,6-bisP is formed from the original sugar, glycolysis is the only metabolic pathway available
      (2) The PFK reaction is highly exergonic and is essentially irreversible
      (3) Molecules that inhibit PFK are ATP (an ultimate product of glycolysis), citrate (an early reactant in the subsequent TCA cycle and a molecule used to synthesize other macromolecules), and certain fatty acids (which also can be oxidized to produce energy); these molecules all help meet the cell's energy requirements; if they are present in abundance, glycolysis will slow down
      (4) An increased concentration of hydrogen ions (indicating a low blood pH) inhibits PFK; slowing glycolysis decreases production of lactic acid (an end product of anaerobic glycolysis) and prevents further lowering of blood pH
      (5) AMP and ADP activate PFK; a increase in AMP or ADP signifies a drop in the cellular energy charge and a need for ATP; thus glycolysis increases to generate more ATP
      (6) Fructose 2,6-bisphosphate activates PFK in liver; it is produced from fructose 6-P using the enzyme phosphofructokinase 2 (PFK 2)
2. The second series of reactions in glycolysis, called the *splitting stage,* splits the six-carbon compound fructose 1,6-bisphosphate into two three-carbon compounds (ultimately two molecules of glyceraldehyde 3-phosphate)
   a. In *Step 4,* the enzyme *aldolase* splits fructose 1,6-bisphosphate into glyceraldehyde 3-phosphate (glyceraldehyde 3-P) and dihydroxyacetone phosphate (dihydroxyacetone-P)
   b. Glyceraldehyde 3-P proceeds directly through glycolysis, but dihydroxyacetone-P does not
   c. In *Step 5,* dihydroxyacetone-P is converted to its isomer, glyceraldehyde 3-P; *triose phosphate isomerase* catalyzes this reversible isomerization
   d. At the end of Step 5, one molecule of glucose has been converted to two molecules of glyceraldehyde 3-P, both of which proceed through the oxidoreduction-phosphorylation stage of glycolysis
3. The *oxidoreduction-phosphorylation stage* converts the two molecules of glyceraldehyde 3-P to two molecules of pyruvate (or lactate, under anaerobic conditions), and produces four molecules of ATP
   a. In *Step 6,* glyceraldehyde 3-P is oxidized and phosphorylated to 1,3-bisphosphoglycerate; this reversible reaction catalyzed by glyceraldehyde 3-P *dehydrogenase* requires the cofactor NAD^+, which is subsequently reduced to $NADH + H^+$

Glycolytic Pathway

The multistep process of glycolysis converts one molecule of glucose to two molecules of pyruvate; the conversion of pyruvate to lactate generally occurs only in anaerobic conditions.

Priming Stage

Step 1. Glucose →(ATP → ADP, hexokinase)→ Glucose 6-phosphate

Step 2. Glucose 6-phosphate →(phosphoglucose isomerase)→ Fructose 6-phosphate

Step 3. Fructose 6-phosphate →(ATP → ADP, phosphofructokinase)→ Fructose 1,6-biphosphate

Splitting Stage

Step 4. Fructose 1,6-biphosphate →(aldolase)→ Dihydroxyacetone phosphate + Glyceraldehyde 3-phosphate

Step 5. Dihydroxyacetone phosphate ⇌(triose phosphate isomerase)⇌ Glyceraldehyde 3-phosphate

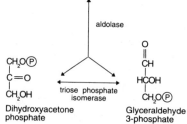

Ⓟ represents a phosphoryl group ($-\overset{\overset{O}{\|}}{\underset{\underset{O^-}{|}}{P}}-O^-$), ~Ⓟ represents a high energy bond

Glycolytic Pathway *(continued)*

Oxidoreduction-Phosphorylation Stage

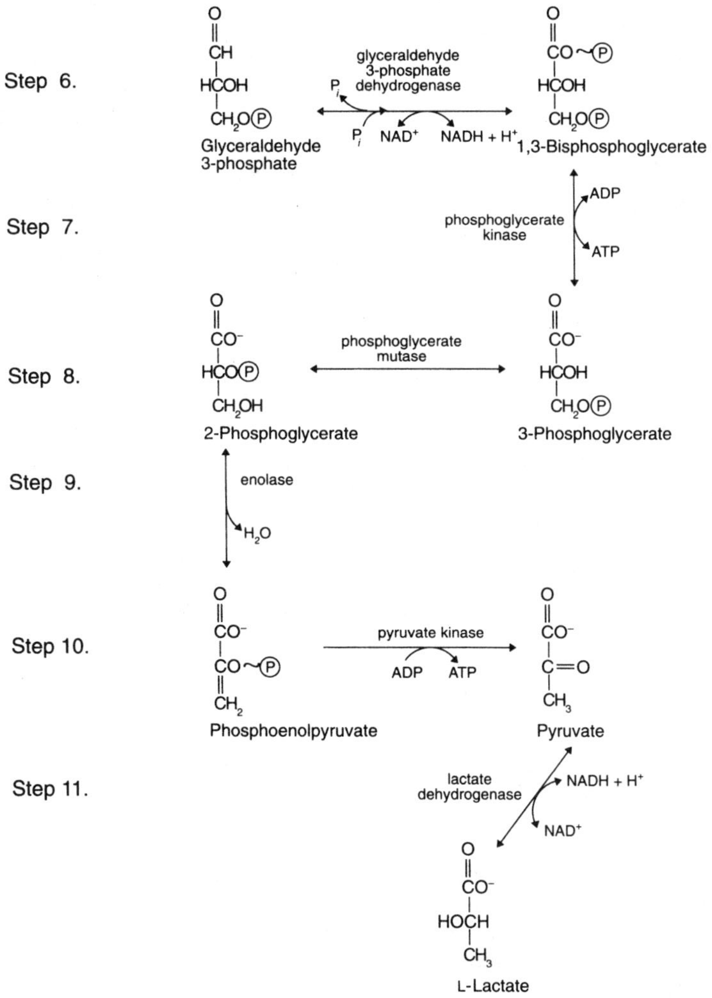

Step 6.

Glyceraldehyde 3-phosphate ⇌ (glyceraldehyde 3-phosphate dehydrogenase, P_i, NAD^+, $NADH + H^+$) 1,3-Bisphosphoglycerate

Step 7.

phosphoglycerate kinase; ADP → ATP

Step 8.

2-Phosphoglycerate ⇌ (phosphoglycerate mutase) 3-Phosphoglycerate

Step 9.

enolase; H_2O

Step 10.

Phosphoenolpyruvate → (pyruvate kinase, ADP → ATP) Pyruvate

Step 11.

lactate dehydrogenase; $NADH + H^+$ → NAD^+; L-Lactate

b. Inorganic phosphate from phosphoric acid forms a high-energy phosphate bond at C-1 of glyceraldehyde 3-P; this high-energy phosphoryl group is a mixed anhydride of carboxylic and phosphoric acid

c. In *Step 7, phosphoglycerate kinase* catalyzes the reversible transfer of the high-energy phosphoryl group from 1,3-bisphosphoglycerate to ADP

d. Because each molecule of glucose produces two molecules of glyceraldehyde 3-P, two ATP molecules are produced in Step 7 from each glucose molecule entering glycolysis

e. Step 7 generates the first pair of ATP molecules; however, because both Steps 1 and 3 use one molecule of ATP, the *net* generation of ATP is zero at this point
f. In *Step 8, phosphoglyceratemutase* catalyzes the reversible rearrangement of 3-phosphoglycerate to 2-phosphoglycerate
g. A diphosphorylated compound, 2,3-bisphosphoglycerate (2,3-BPG) is formed as an intermediate of Step 8; this same 2,3-BPG is an important allosteric effector of the reversible association of oxygen with hemoglobin (see Chapter 4, Protein Function and Metabolism)
h. In *Step 9, enolase* catalyzes the reversible dehydration of 2-phosphoglycerate to form phosphoenolpyruvate (PEP) and water; the high-energy phosphate bond of PEP results from the formation of a double bond ("ene") adjacent to a hydroxyl group ("ol"), which increases the group transfer potential of the phosphoryl group
i. In *Step 10,* the third irreversible reaction of glycolysis, *pyruvate kinase (PK)* catalyzes the transfer of the high energy phosphoryl group from phosphoenolpyruvate to ADP, producing pyruvate and another ATP molecule
j. This final ATP-producing step yields two molecules of ATP for each glucose molecule, because each molecule of glucose yields two molecules of glyceraldehyde 3-P; accounting for ATP molecules used in Steps 1 and 3, and produced in Step 7, Step 10 yields a net generation of two ATP molecules in the glycolytic metabolism of one glucose molecule
k. Pyruvate kinase is an allosteric enzyme affected by various molecules
   (1) In the liver, high concentration of ATP inhibits PK; because pyruvate is transaminated to the amino acid alanine (see Chapter 4, Protein Function and Metabolism), abundant alanine inhibits PK; PK also is inhibited by covalently bonded phosphoryl groups (covalent modification)
   (2) Fructose 1,6-bis P activates PK because a high fructose 1,6-bis P concentration signals an accumulation of early stage glycolytic products and a need for PK to catalyze formation of pyruvate at a faster rate
l. In *Step 11,* pyruvate is reduced to lactate through the action of the enzyme *lactate dehydrogenase;* this part of the glycolytic path, called anaerobic homolactic fermentation, is used by muscle cells working under anaerobic conditions (oxygen depletion accompanied by high energy demands)
m. $NADH + H^+$, the required cofactor for lactate dehydrogenase, is oxidized to NAD^+; the NAD^+ is reused in the glyceraldehyde 3-P dehydrogenase reaction (Step 6 of glycolysis), thus maintaining glycolysis in muscles under anaerobic conditions
n. Under aerobic conditions (when oxygen is sufficient), pyruvate does not proceed through Step 11 of glycolysis; instead it is decarboxylated to produce carbon dioxide and acetate; acetate is complexed with CoA to form acetyl CoA and enters the TCA cycle (see section III. in this chapter)

4. In yeast and certain bacteria, pyruvate is converted to ethanol and CO_2 by anaerobic alcoholic fermentation
   a. Pyruvate is first decarboxylated to acetaldehyde and CO_2 in a reaction catalyzed by pyruvate decarboxylase, an enzyme not found in animals
   b. The enzyme alcohol dehydrogenase subsequently catalyzes the reduction of acetaldehyde to ethanol and the oxidation of NADH + H^+ to NAD^+
   c. Note that both anaerobic homolactic fermentation (Step 11 of anaerobic glycolysis) and anaerobic alcoholic fermentation generate NAD^+ for the continuance of glycolysis under anaerobic conditions

III. The Tricarboxylic Acid (TCA) Cycle

A. General information
1. The TCA cycle (also called the citric acid cycle or Krebs cycle) is a series of enzyme-catalyzed reactions occurring only in the presence of oxygen; it occurs in the mitochondria of eukaryotic organisms or the plasma membrane of prokaryotic organisms; individual two-carbon units are carried into the TCA cycle as molecules of acetyl CoA and leave the cycle as CO_2
2. At the end of the TCA cycle, each glucose molecule that entered glycolysis has been oxidized completely to six CO_2 molecules; through glycolysis, each glucose molecule is converted to two molecules of pyruvate; through oxidative decarboxylation (the intervening step between glycolysis and the TCA cycle) two pyruvate molecules yield two CO_2 molecules; through the TCA cycle, two acetyl CoA molecules yield four CO_2 molecules
   a. The enzyme *pyruvate dehydrogenase* catalyzes the oxidative decarboxylation of pyruvate to CO_2 and acetyl CoA; NAD^+ is reduced to $NADH^+ + H^+$; this irreversible reaction occurs in the matrix of mitochondria
   b. Pyruvate dehydrogenase is a complex enzyme that contains three catalytic subunits: pyruvate dehydrogenase, dihydrolipoyl transacetylase, and dihydrolipoyl dehydrogenase
   c. In addition to its requirements for CoA and NAD^+, pyruvate dehydrogenase requires the cofactors thiamine pyrophosphate (TPP), lipoic acid, and FAD
   d. The activity of pyruvate dehydrogenase is inhibited by elevated concentrations of NADH, acetyl CoA, and GTP, and is activated by AMP (an increase in AMP indicates low energy charge and a need for the TCA cycle to generate energy)
   e. In mammals, pyruvate dehydrogenase is inhibited via phosphorylation (covalent modification); the phosphorylated form is inactive
3. The TCA cycle is the final common path for the oxidation of amino acids, fatty acids, and carbohydrates
   a. In the beginning of the cycle, acetyl CoA forms a complex with oxaloacetate
   b. At the completion of the cycle, oxaloacetate is regenerated to complex with a new molecule of acetyl CoA

c. In three different oxidation-reduction reactions of the cycle, NAD^+ is reduced to $NADH + H^+$; in one oxidation-reduction reaction FAD is reduced to $FADH_2$
d. The electrons carried by NADH and $FADH_2$ ultimately enter the respiratory chain
4. Because the the TCA pathway is cyclic, it has no ultimate products; TCA intermediates serve as building blocks for other biosynthetic reactions (see *Tricarboxylic Acid [TCA] Cycle*)

B. Reactions of the TCA Cycle

1. In *Step 1, citrate synthase* catalyzes the condensation of acetyl CoA and oxaloacetate to form citrate; citrate performs other roles besides energy generation in the TCA cycle; its carbons are used as building blocks for such biosynthetic reactions as gluconeogenesis, and it is an allosteric regulator for other enzymes, such as phosphofructokinase, in glycolysis
2. In *Step 2*, the enzyme *aconitase* catalyzes the isomerization of citrate to isocitrate via the enzyme-bound intermediate, *cis*-aconitate
3. In *Step 3, isocitrate dehydrogenase* catalyzes the oxidative decarboxylation of isocitrate to α-ketoglutarate and CO_2; this reaction is the rate-limiting step of the TCA cycle
   a. Oxalosuccinate is formed as an intermediate and the coenzyme NAD^+ is reduced to $NADH + H^+$
   b. The reaction is controlled by the concentration of NADH and ATP, which inhibit the activity of isocitrate dehydrogenase; ADP activates isocitrate dehydrogenase
   c. The first molecules of CO_2 and $NADH + H^+$ generated by the TCA cycle are produced by this reaction
4. *Step 4* is the second oxidative decarboxylation of the TCA cycle; α-ketoglutarate is oxidatively decarboxylated to CO_2 and succinate, which then complexes with CoA to yield succinyl CoA
   a. The multienzyme complex *α-ketoglutarate dehydrogenase* catalyzes this reaction; it is inhibited by increased levels of its products—succinyl CoA, NADH, and ATP
   b. Like pyruvate dehydrogenase, α-ketoglutarate dehydrogenase is composed of three subunits (α-ketoglutarate dehydrogenase, dihydrolipoyl transuccinylase, and dihydrolipoyl dehydrogenase) and requires the cofactors CoA, NAD^+, TPP, lipoic acid, and FAD
   c. The second and final CO_2 and the second $NADH + H^+$ of the TCA cycle are produced by this reaction
5. In *Step 5*, the high-energy thioester bond of succinyl CoA is hydrolyzed to form succinate and CoA; in mammals, the energy released from hydrolysis of the thioester bond is used for substrate level phosphorylation (conservation of the high amount of energy in the substrate while phosphorylating another compound) of GDP to GTP; *succinyl CoA synthetase* (also called succinate thiokinase) catalyzes this reaction
6. In *Step 6, succinate dehydrogenase* catalyzes the oxidation of succinate to fumarate

Tricarboxylic Acid (TCA) Cycle

The TCA cycle is the final common path for the oxidation of the carbons of amino acids, fatty acids, and carbohydrates.

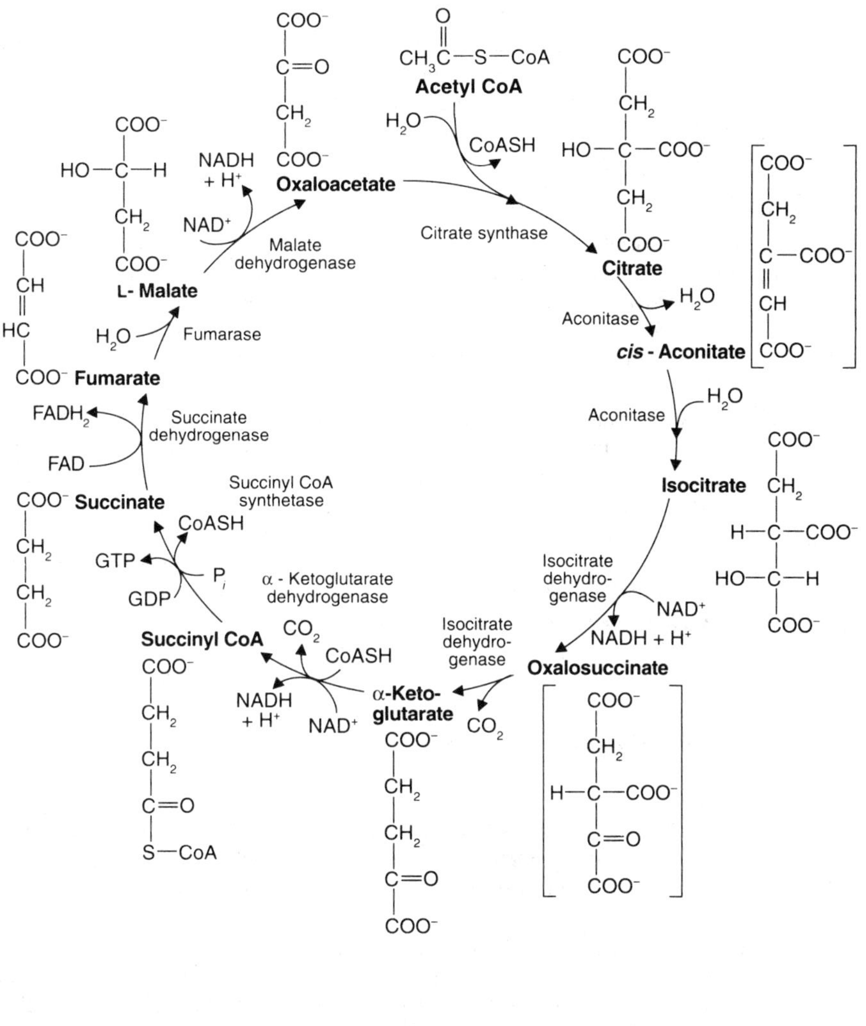

Oxaloacetate
$$\begin{array}{c} COO^- \\ | \\ C=O \\ | \\ CH_2 \\ | \\ COO^- \end{array}$$

Acetyl CoA: $CH_3\text{-}C(=O)\text{-}S\text{-}CoA$

L-Malate
$$\begin{array}{c} COO^- \\ | \\ HO\text{-}C\text{-}H \\ | \\ CH_2 \\ | \\ COO^- \end{array}$$

Malate dehydrogenase: NADH + H⁺ ← NAD⁺

Citrate synthase: H₂O, CoASH

Citrate
$$\begin{array}{c} COO^- \\ | \\ CH_2 \\ | \\ HO\text{-}C\text{-}COO^- \\ | \\ CH_2 \\ | \\ COO^- \end{array}$$

Aconitase → H₂O

cis-Aconitate
$$\left[\begin{array}{c} COO^- \\ | \\ CH_2 \\ | \\ C\text{-}COO^- \\ \| \\ CH \\ | \\ COO^- \end{array}\right]$$

Aconitase ← H₂O

Isocitrate
$$\begin{array}{c} COO^- \\ | \\ CH_2 \\ | \\ H\text{-}C\text{-}COO^- \\ | \\ HO\text{-}C\text{-}H \\ | \\ COO^- \end{array}$$

Isocitrate dehydrogenase: NAD⁺ → NADH + H⁺

Oxalosuccinate
$$\left[\begin{array}{c} COO^- \\ | \\ CH_2 \\ | \\ H\text{-}C\text{-}COO^- \\ | \\ C=O \\ | \\ COO^- \end{array}\right]$$

Isocitrate dehydrogenase → CO₂

Fumarate
$$\begin{array}{c} COO^- \\ | \\ CH \\ \| \\ HC \\ | \\ COO^- \end{array}$$

Fumarase ← H₂O

Succinate dehydrogenase: FAD → FADH₂

Succinate
$$\begin{array}{c} COO^- \\ | \\ CH_2 \\ | \\ CH_2 \\ | \\ COO^- \end{array}$$

Succinyl CoA synthetase: CoASH, GTP ← GDP + P_i

Succinyl CoA
$$\begin{array}{c} COO^- \\ | \\ CH_2 \\ | \\ CH_2 \\ | \\ C=O \\ | \\ S\text{-}CoA \end{array}$$

α-Ketoglutarate dehydrogenase: NADH + H⁺ ← NAD⁺, CoASH, CO₂

α-Ketoglutarate
$$\begin{array}{c} COO^- \\ | \\ CH_2 \\ | \\ CH_2 \\ | \\ C=O \\ | \\ COO^- \end{array}$$

a. Succinate dehydrogenase is an integral protein of the inner mitochondrial membrane; all other enzymes of the TCA cycle are free-moving proteins in the mitochondrial matrix
b. Like all other dehydrogenases, the enzyme requires a cofactor, but uses FAD (instead of NAD⁺), which is reduced to FADH₂; as with NADH⁺, FADH₂ is ultimately oxidized in the respiratory chain to produce ATP

7. In *Step 7,* the enzyme *fumarase* catalyzes the hydration of the fumarate double bond to form malate
8. In *Step 8, malate dehydrogenase* catalyzes the oxidation of malate to oxaloacetate, completing the cycle; NAD^+, the required cofactor, is reduced yielding the third and last $NADH + H^+$ of the TCA cycle
9. In terms of stored chemical energy, one glucose molecule entering the TCA cycle as two molecules of acetyl CoA yields two molecules of GTP (from the hydrolysis of two molecules of succinyl CoA), six molecules of $NADH + H^+$ (from the oxidation of two molecules each of isocitrate, α-ketoglutarate, and malate) and two molecules of $FADH_2$ (from the oxidation of two molecules of succinate)

C. The glyoxylate cycle
1. The glyoxylate cycle, which resembles the TCA cycle, enables plants and bacteria to use acetate or other compounds that can be converted to acetyl CoA to generate energy and provide biosynthetic products
2. The glyoxylate cycle differs from the TCA cycle in several ways
    a. Two acetyl units enter the glyoxylate cycle together; acetyl units enter the TCA cycle one at a time
    b. One acetyl unit forms isocitrate, just as in the normal TCA cycle; however rather than forming α-ketoglutarate and subsequently succinyl CoA, this isocitrate molecule is cleaved to yield succinate, the product of the cycle, and glyoxylate (bypassing the two decarboxylation steps of the TCA cycle)
    c. Glyoxylate reacts with the second molecule of acetyl CoA (see a. above) to form malate, which is then oxidized to oxaloacetate
    d. Plants and bacteria use the newly formed oxaloacetate to make glucose through a process called gluconeogenesis

IV. Oxidative Phosphorylation in the Respiratory Chain

A. General information
1. Oxidative phosphorylation is a process whereby ATP is produced as electrons are shuttled through the energy-generating components of the respiratory chain (also called the electron transport chain) to molecular oxygen (O_2), forming water
    a. Electrons are donated from molecules of NADH and $FADH_2$ produced in the complete oxidation of glucose to CO_2 and water (remember that one glucose molecule enters the TCA cycle as two acetyl CoA molecules); NADH and $FADH_2$ are oxidized in oxidative phosphorylation
        (1) Two molecules of NADH result from glycolysis of one molecule of glucose
        (2) Two molecules of NADH result from the decarboxylation of two pyruvate molecules to acetyl CoA in the reaction linking glycolysis to the TCA cycle
        (3) Six molecules of NADH and two molecule of $FADH_2$ result from the TCA cycle oxidation of two acetyl CoA molecules
    b. The basic oxidative phosphorylation reaction is simply: $ADP + P_i \rightarrow ATP$

c. The oxidation-reduction reactions of oxidative phosphorylation are a separate biochemical event linked to the formation of ATP; the two reactions are linked by a ***proton-motive force***
   d. In the oxidation-reduction portion of oxidative phosphorylation, each NADH + H$^+$ and FADH$_2$ donates a pair of electrons to a series of intermediate compounds that eventually carry the donated electrons to oxygen, which accepts the electrons and forms water
   e. The result is that glucose is completely oxidized to CO$_2$ by the end of the TCA cycle and water by the end of oxidative phosphorylation
  2. During oxidative phosphorylation, the oxidation of each NADH produces three molecules of ATP; the oxidation of each FADH$_2$ produces two molecules of ATP
  3. Oxidative phosphorylation is the major source of ATP for aerobic organisms
  4. In eukaryotes, oxidative phosphorylation occurs in the mitochondrial inner membrane; in prokaryotes, it occurs in the cytoplasmic membrane

B. **Energetics of oxidative phosphorylation**
  1. As electrons are transferred sequentially from NADH and FADH$_2$ to O$_2$ (the final electron acceptor), protons (H$^+$) are translocated ("pumped") out of the mitochondrial matrix into the intermembrane space between the inner mitochondrial membrane and the outer mitochondrial membrane
  2. Pumping of H$^+$ from the mitochondrial matrix into the intermembrane space results in both a pH gradient and generation of an electrical potential across the inner mitochondrial membrane; this concentration gradient and electrical potential tends to drive H$^+$ back through the inner mitochondrial membrane and into the mitochondrial matrix
  3. The movement of H$^+$ back into the mitochondrial matrix is linked to an enzyme complex called ATP synthetase or ATPase; as the protons move through this enzyme complex, they foster the synthesis and release of ATP from ADP and P$_i$
  4. Complete oxidation of one molecule of glucose (via glycolysis, the TCA cycle, and the respiratory chain) to CO$_2$ and water yields either 38 or 36 molecules of ATP, depending on how the NADH produced in glycolysis is transported into the mitochondria to participate in the respiratory chain
     a. Recall that in eukaryotes the pyruvate dehydrogenase reaction, the TCA cycle, and oxidative phosphorylation all occur in the mitochondria; the NADH and FADH$_2$ generated is immediately available for oxidation through the respiratory chain
     b. Glycolysis occurs in the cytosol; NADH generated in glycolysis must enter the mitochondria for oxidative phosphorylation
     c. Two different shuttle mechanisms move glycolytic NADH from the cytosol to mitochondria: the glycerol-3-phosphate shuttle and the malate-aspartate shuttle (see section IV. C. 9. in this chapter)
     d. The glycerol-3-phosphate shuttle, which is active in muscle cells, produces two molecules of ATP for every glycolytic NADH moved into the mitochondria
     e. The malate-aspartate shuttle, active in liver and heart cells, produces three molecules of ATP for every NADH moved into the mitochondria from the cytosol

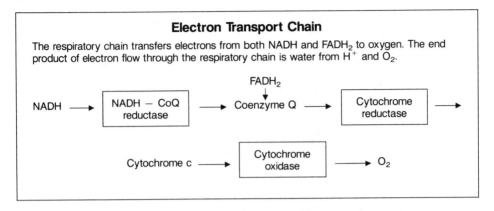

5. Thirty-two of the 36 (or 34 of the 38) ATP molecules produced from the complete oxidation of one molecule of glucose (via glycolysis, the TCA cycle, and the respiratory chain) to CO_2 and water are produced by oxidative phosphorylation. When reviewing the following sites of ATP production, recall that one molecule of glucose is split into two three-carbon units, doubling the number of NADH and $FADH_2$ produced.
   a. The two molecules of NADH produced in glycolysis yield either 2 ATP molecules each (when transported by the glycerol-3-phosphate shuttle) or 3 ATP molecules each when transported by the malate-aspartate shuttle, thus generating either 4 or 6 molecules of ATP, respectively
   b. The two molecules of NADH produced from the decarboxylation of pyruvate in the reaction linking glycolysis to the TCA cycle yield 3 ATP
   c. The six molecules of NADH produced in the TCA cycle yield 3 ATP molecules each, thus generating 18 molecules of ATP
   d. The two molecules of $FADH_2$ produced in the TCA cycle yield 2 ATP molecules, thus generating 4 molecules of ATP
   e. The sum of the above reactions is 32 or 34 ATP molecules; the remaining 4 molecules of ATP are generated directly in glycolysis (2 ATP) and in the TCA cycle (2 GTP converted to 2 ATP)

C. **Respiratory chain components**
   1. The respiratory chain transfers electrons and is composed of many different complexes, including proteins that act as enzymes in electron transfer as well as specialized proteins called *cytochromes;* a cytochrome is an electron-transferring protein with a heme prosthetic group
   2. As electrons flow from NADH to O_2, they are transferred sequentially to two intermediate compounds, coenzyme Q (CoQ) and cytochrome c (Cyt c), and then ultimately to molecular oxygen from Cyt c (see *Electron Transport Chain*)
   3. Three different large enzyme complexes catalyze the flow of electrons between CoQ, Cyt c, and oxygen
      a. *NADH-CoQ reductase* catalyzes electron flow between NADH and CoQ
      b. *Cytochrome reductase* catalyzes electron flow between CoQ and Cyt c

c. *Cytochrome oxidase* catalyzes electron flow between Cyt c and O_2 to form H_2O; oxygen has a high affinity for electrons and is relatively unreactive, making it a good choice for a terminal electron acceptor
4. The three enzyme complexes are the sites of both H^+ pumping and movement of electrons; the actual groups that carry electrons in the respiratory chain enzymes are flavins, iron and sulfur atoms in proteins (iron-sulfur clusters), heme groups, and copper ions (Cu^{2+})
5. Electrons in the respiratory chain are first transferred from NADH to CoQ via the NADH-CoQ reductase complex, the first of the three enzymatic complexes that both transfer electrons and act as proton pumps
   a. The NADH-CoQ reductase complex first accepts pairs of electrons from NADH at one of its prosthetic groups, flavin mononucleotide (FMN), reducing this prosthetic group to $FMNH_2$; if only one electron is transferred, an unstable intermediate of CoQ, called a semiquinone (QH), is formed
   b. Electrons are next transferred from $FMNH_2$ to a second prosthetic group of NADH-CoQ reductase, the iron-sulfur proteins, also called nonheme iron proteins because the iron atom is bound to sulfur instead of heme
   c. Electrons are then transferred from the iron-sulfur prosthetic group to the oxidized form of CoQ (also called ubiquinone); CoQ accepts electrons to form the reduced compound $CoQH_2$ (also called ubiquinol)
   d. CoQ is also the site at which electrons from $FADH_2$, produced in the TCA cycle, enter the respiratory chain, thus bypassing the NADH-CoQ reductase complex
6. The second enzyme complex of the respiratory chain, the cytochrome reductase complex, transfers electrons from ubiquinol to Cyt c, the second of the proton pumps
   a. Cytochrome reductase contains an iron-sulfur cluster and other proteins, plus two types of cytochromes, cytochrome b (Cyt b) and cytochrome c_1 (Cyt c_1)
   b. Only one electron at a time is transferred from ubiquinol, generating a semiquinone (QH)
   c. The first electron is transferred to the iron-sulfur cluster in cytochrome reductase, then to Cyt c_1, then to Cyt c
   d. The second electron is transferred to heme groups in Cyt b; Cyt b donates this electron to QH, reducing it to $CoQH_2$; Cyt b is an electron recycling device that enables ubiquinol, a two-electron carrier, to transfer one electron at a time to an iron-sulfur cluster of cytochrome reductase, a single-electron carrier
7. The third enzyme complex in the respiratory chain, the cytochrome oxidase complex, transfers electrons from Cyt c to O_2 to form H_2O; four electrons are needed to completely reduce one molecule of O_2 to two molecules of H_2O; this is the last of the three proton pumps
   a. Cytochrome oxidase contains cytochromes with two different heme groups (Cyt a and Cyt a_3) and two copper ions (Cu_A and Cu_B)
   b. Cyt c donates its electrons to the Cyt a–Cu_A complex, then one electron at a time is transferred to the Cyt a_3–Cu_B complex, then to O_2, thus producing two H_2O molecules

8. The movement of electrons through all three enzyme complexes of the respiratory chain results in pumping of H⁺ from the matrix of the mitochondria to the intermembrane space; the resulting difference in electrical potential and pH between the matrix and the intermembrane space drives the production of ATP through a mechanism called the proton-motive force
9. The NADH molecules produced in glycolysis are the only glycolytic NADH molecules produced outside the mitchondria; to enter the respiratory chain, the NADH molecules must first traverse the mitochondrial membrane through two different shuttle mechanisms
    a. Actually, neither shuttle mechanism transfers the NADH molecule across the inner mitochondrial membrane; rather, the shuttles transfer the electrons from NADH
    b. The two shuttle mechanisms yield different amounts of ATP; the glycerol phosphate shuttle yields two molecules of ATP per NADH; the malate-aspartate shuttle yields three molecules of ATP per NADH
    c. The glycerol-phosphate shuttle, active in muscle cells, begins when a dihydroxyacetone phosphate molecule in the cytosol is reduced to glycerol-3-phosphate by accepting electrons from cytosolic NADH that results from glycolysis
        (1) The glycerol-3-phosphate molecule diffuses into the mitochondria, delivers its electrons to FAD to form $FADH_2$, and is simultaneously oxidized back to dihydroxyacetone phosphate
        (2) The recreated dihydroxyacetone phosphate then exits the mitochondria, thus completing the shuttle
        (3) $FADH_2$ produced within the mitochondria enters the respiratory chain at CoQ, bypassing the NADH–CoQ complex, which is a site of a proton pump; thus, only two ATP molecules are produced
    d. The malate-aspartate shuttle, active in heart and liver cells, functions because malate can cross mitochondrial membrane while oxaloacetate cannot; in the cytosol, an oxaloacetate molecule accepts electrons from NADH produced during glycolysis and is reduced to malate
        (1) The malate molecule crosses into the mitochondrion and is oxidized to oxaloacetate; oxaloacetate is transaminated to aspartate with the simultaneous production of NADH + H⁺ from NAD⁺
        (2) Aspartate then leaves the mitochondria and diffuses into the cytosol, where it is converted back to oxaloacetate to complete the shuttle
        (3) The NADH generated within the mitochondria enters the respiratory chain at NADH–CoQ reductase to produce three molecules of ATP

D. **Generation of ATP by proton-motive force**
    1. The flow of electrons from NADH to O_2 is a thermodynamically favorable reaction; the standard free energy change ($\Delta G^{\circ\prime}$) for the reaction $\frac{1}{2} O_2 + NADH + H^+ \rightleftharpoons H_2O + NAD$ can be calculated
        a. For this reaction, which is the overall result of the electron transport of NADH electrons, $\Delta G^{\circ\prime} = -52.6$ kcal/mol
        b. The free energy expended during the synthesis of one molecule of ATP from ADP is +7.3 kcal/mol ($\Delta G^{\circ\prime} = +7.3$ kcal/mol)

c. The exergonic formation of water in the respiratory chain provides energy for the endergonic formation of ATP
2. Recall that a mitochondrion has two primary membranes, an outer membrane and an inner membrane that invaginates into a central space called the matrix; the space between the inner membrane and the outer membrane is called the intermembrane space
3. During the synthesis of ATP, protons flow from the intermembrane space back into the matrix through special ion channels; these channels are a structural feature of the enzyme complex that synthesizes ATP, called ATPase or ATP synthase
4. ATPase, located in the inner mitochondrial membrane, is composed of three different protein subunits: F_1, F_0, and a stalk protein
   a. F_1, composed of five polypeptides and located on the matrix side of the inner mitochondrial membrane, catalyzes the reaction ADP + P_i ⇌ ATP
   b. F_0, composed of four polypeptides and spanning the inner mitochondrial membrane, is the membrane channel through which H^+ passes
   c. The stalk protein links F_0 to F_1
   d. The ADP concentration is a major factor in determining the rate of oxidative phosphorylation; a low ADP concentration increases the rate of oxidative phosphorylation
   e. Electrons usually are transferred in the respiratory chain only when ATP is needed; this is called respiratory control
5. The *chemiosmotic hypothesis* describes how the proton gradient between the mitochondrial matrix and the intermembrane space results in production of ATP
   a. In the respiratory chain, the transfer of electrons from one complex to the next leads to H^+ pumping from the mitochondrial matrix across the inner mitochondrial membrane into the intermembrane space
   b. H^+ pumping results in a buildup of H^+ and a positive electrical potential on the outer side of the inner mitochondrial membrane
   c. The pH gradient and membrane electrical potential generate a proton-motive force (PMF), which does not actually form ATP, but rather allows ATP to be released from ATPase, the enzyme that catalyzes its synthesis
   d. The chemisomotic theory does not describe how the movement of H^+ through ATPase causes the generation of ATP; the theory merely describes how ATPase forms a channel for the redistribution of H^+ while somehow simultaneously generating ATP
   e. A pH gradient is generated at each of the three electron transfer complexes where H^+ pumping occurs (NADH-CoQ reductase, cytochrome reductase, and cytochrome oxidase); one molecule of ATP is synthesized at each site of H^+ pumping
   f. The P:O ratio, the number of molecules of inorganic phosphate incorporated into ADP per atom of oxygen consumed, is an index of oxidative phosphorylation
   g. Because electrons from NADH are transferred through all three respiratory complexes at which H^+ pumping can occur, the oxidation of NADH has a P:O ratio of 3; because electrons from $FADH_2$ enter the respiratory chain at CoQ, thereby bypassing NADH-Q reductase, the oxidation of $FADH_2$ has a P:O ratio of 2

E. Respiratory chain inhibitors and uncouplers

1. The three H^+ pumping sites of the respiratory chain can be blocked by various inhibitors
   a. NADH-CoQ reductase is inhibited by rotenone (a fish poison) and amytal (a barbiturate); these compounds do not inhibit the oxidation of $FADH_2$, because $FADH_2$ enters the respiratory chain after this inhibition site
   b. Cytochrome reductase is inhibited by the antibiotic antimycin A, which inhibits electron transfer at the Cyt b level
   c. Antimycin A inhibition can be overcome by adding ascorbate (vitamin C, a reducing agent), which directly reduces Cyt c, a step farther along the chain
   d. Cytochrome oxidase is inhibited by cyanide ($-CN-$), azide ($-N_3$), and carbon monoxide (CO); cyanide and azide complex with cytochrome heme groups in cytochromes a and a_3 and prevent electron transfer; carbon monoxide binds to cytochrome oxidase
2. Continued exposure to respiratory chain inhibitors leads to death of the organism from insufficient energy production
3. Because oxidation of NADH and phosphorylation of ATP are separate biochemical reactions linked only by a PMF, substances that carry H^+ across the mitochondrial inner membrane dissipate the pH gradient required for ATP synthesis
   a. These substances are called uncouplers because they uncouple or dissociate the oxidation of NADH (which still can continue) from the phosphorylation of ADP (which they inhibit)
   b. Examples of uncouplers, substances that increase the H^+ permeability of the inner mitochondrial membrane, are 2,4-dinitrophenol and dicumarol
   c. Uncoupling of NADH oxidation from ADP phosphorylation is a mechanism to generate heat in hibernating and newborn animals, including humans
4. Oligomycin (an antibiotic) inhibits mitochondrial ATPase directly; because oxidation is coupled to phosphorylation, oligomycin inhibits both ATP synthesis and electron transport

---

Study Activities

1. Define the cellular sites for glycolysis, the TCA cycle, and the respiratory chain.
2. Account for the high-energy molecules produced and consumed in glycolysis, the TCA cycle, and oxidative phosphorylation in the respiratory chain.
3. Explain how the change in Gibbs free energy associated with a reaction predicts whether or not the reaction is spontaneous.
4. Explain the function, mechanisms, and differences between the glycerol-3-phosphate shuttle and the malate-aspartate shuttle.
5. In oxidative phosphorylation, describe what is oxidized, what is phosphorylated, and how both reactions are accomplished.
6. Distinguish between an inhibitor and an uncoupler of the respiratory chain.
7. Define the terms listed in bold-faced type throughout this chapter.

4

Protein Function and Metabolism

Objectives

After studying this chapter, the reader should be able to:
- Describe the structural and functional aspects of the reversible binding of oxygen to hemoglobin.
- Describe enzymes in terms of their composition, binding sites, nonprotein components, and reaction with inhibitors.
- Discuss the information required to determine the rate of a particular enzyme-catalyzed reaction.
- Describe the three-dimensional structure and the function of collagen.
- Describe the role of actin and myosin in skeletal muscle contraction.
- Relate the structure of ferritin to its role as a storage protein.
- Describe the reasons for antibody diversity and how antibody diversity is achieved.
- Discuss how and why proteins may be complexed with carbohydrates, lipids, and nucleic acids.
- Discuss the fates of both the amine group and the carbon skeleton in the catabolism of amino acids.

I. Diversity of Protein Functions

A. General information
1. Proteins are involved in some way in all the biochemical reactions that make life possible; each species possesses proteins that are distinct from those of all other species
2. Proteins *transport substances;* for example, the protein hemoglobin transports both oxygen and carbon dioxide in the blood
3. Proteins *catalyze biochemical reactions* (a catalyst is a substance that increases the rate of a chemical reaction without a net chemical change to itself); the majority of enzymes (biological catalysts) are proteins
4. Proteins help *maintain structure;* for example, the protein collagen constitutes part of fibrous connective tissues in skin, bone, tendon, cartilage, blood vessels, and teeth
5. Proteins *facilitate movement;* for example, the proteins actin and myosin mediate muscle contraction
6. Proteins *serve as storage compounds;* for example, the protein ferritin stores iron in the body

> **A Molecule of Iron Protoporphyrin IX**
>
> Each of the four polypeptide chains in a molecule of hemoglobin is tightly bound to a prosthestic group called iron protoporphyrin IX.

7. Proteins *protect organisms against foreign invaders,* such as viruses, bacteria, and the toxins they produce; for example, immunoglobulins (antibodies) are proteins that react with and neutralize foreign compounds (antigens)
8. Some proteins *serve as hormones,* the messengers and regulators of metabolism; for example, the protein hormone insulin regulates glucose levels in the blood

B. **Transport role of proteins: Hemoglobin**
   1. Hemoglobin (Hb), an oligomeric protein found in erythrocytes (red blood cells), is composed of four polypeptide chains (subunits), two each of two identical subunits
   2. Hemoglobin is a **globular** protein with considerable α-helical secondary structure; different hemoglobin types vary in subunit composition
      a. Hb A, the major type of adult hemoglobin, is composed of two α and two β polypeptide chains
      b. Hb A_2, comprising approximately 3% of adult hemoglobin, is composed of two α and two δ polypeptide chains
      c. Hb F (fetal hemoglobin), found in fetuses and newborns, is composed of two α and two γ polypeptide chains
   3. Each of the four polypeptide chains in hemoglobin is tightly bound to its own molecule of *heme,* which functions as a **prosthetic group** (a nonprotein portion of a protein molecule that is necessary for its activity)
      a. Heme is a heterocyclic molecule that contains an iron atom in its center; the heme of hemoglobin is called Fe(II) heme, iron protoporphyrin IX, or ferroprotoporphyrin IX (see *A Molecule of Iron Protoporphyrin IX*)

    b. Protoporphyrin IX, the nonmetal portion of the heme molecule, consists of four pyrrole rings linked by methene bridges to form a tetrapyrrole structure
    c. The iron atom bound in the center of the protoporphyrin ring is the actual binding site of oxygen (O_2) in the hemoglobin molecule; each iron atom binds one O_2 molecule
    d. Hemoglobin binds O_2 when iron is in the reduced state (abbreviated as Fe[II], Fe^{2+}, or ferrous); hemoglobin does not bind O_2 when iron is in the oxidized — Fe(III), Fe^{3+}, or ferric — state
    e. Because each of the four polypeptide chains in hemoglobin has its own heme group, one hemoglobin molecule binds four O_2 molecules (the iron in each heme binds one O_2 molecule)
    f. The four heme groups are located in nonpolar crevices near the surface of the hemoglobin molecule, but far apart from each other
4. Hemoglobin *binds* O_2 for transport to cells that require oxygen and then *releases* O_2 at those cells; oxyhemoglobin is hemoglobin bound to O_2; deoxyhemoglobin is hemoglobin without bound O_2
    a. O_2 binding to hemoglobin is cooperative; binding of the first O_2 molecule to the first heme group facilitates binding of subsequent O_2 molecules to the other three heme groups
    b. The cooperative binding of O_2 to hemoglobin demonstrates an **allosteric effect,** an effect in which the binding of one molecule at a specific site on a protein facilitates the binding of subsequent molecules at the distant sites on the protein; in hemoglobin, the binding of the first O_2 molecule facilitates binding of subsequent O_2 molecules, even though the binding sites are distant from each other
    c. O_2 binding to deoxyhemoglobin breaks eight ionic bonds and causes partial disruption of its normal quaternary structure as it becomes oxyhemoglobin
    d. The molecule *2,3-bisphosphoglycerate* (BPG) causes hemoglobin to release O_2 molecules at body cells needing oxygen
        (1) One molecule of BPG binds in the central cavity of a deoxyhemoglobin molecule, stabilizing its quaternary structure and thereby favoring the deoxygenated form of hemoglobin
        (2) Upon reoxygenation, the conformation of hemoglobin changes, making the central cavity of the protein too small for BPG; without BPG, hemoglobin can take up O_2 and form oxyhemoglobin
    e. An *oxygen dissociation curve* is a graphic representation of the saturation of binding sites (heme groups) as a function of the amount of O_2 present; O_2 is measured as the partial pressure of gaseous O_2 (pO_2) in units of torr (see *Oxygen Dissociation Curve for Hemoglobin and Myoglobin,* page 58)
        (1) The pO_2 at which 50% of binding sites are saturated with O_2 is termed the p_{50}
        (2) The p_{50} for different O_2-binding proteins is unique and predicts the affinity of the protein for O_2

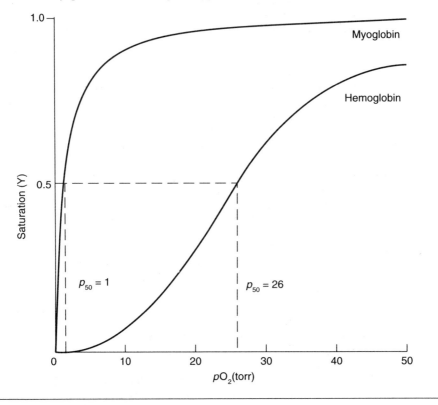

(3) A partial pressure of 26 torr is required to saturate half the binding sites in hemoglobin, but only 1 torr saturates half the binding sites in myoglobin, indicating that myoglobin has a higher affinity for O_2 than hemoglobin

(4) The sigmoidal (S-shaped) oxygen dissociation curve for hemoglobin reflects the cooperative binding of O_2 to hemoglobin; in contrast, the hyperbolic oxygen dissociation curve for myoglobin suggests no cooperative binding

5. The *Bohr effect* describes the relationships between the binding of O_2, CO_2, and H^+ to hemoglobin

   a. Metabolically active tissues require O_2 and produce CO_2 and H^+ as by-products; hemoglobin releases O_2 to these tissues and picks up CO_2 and H^+

   b. Accumulation of CO_2 and H^+ in the hemoglobin molecule decreases its affinity for O_2

c. CO_2 generates H^+ in erythrocytes by the reaction $CO_2 + H_2O \rightleftarrows H^+ + HCO_3^-$, catalyzed by carbonic anhydrase (see Chapter 1, Basics of Biochemistry)
   d. Hemoglobin transports CO_2 to the lungs, which eliminate CO_2 in expired air; H^+ is handled in two ways
      (1) Hemoglobin's four N-terminal amine groups and the imidazole groups of its histidine residues take up H^+; these groups are responsible for hemoglobin's blood-buffering capacity
      (2) Deoxyhemoglobin can take up H^+ without a resultant change in pH, if the extent to which H^+ is generated from CO_2 is just sufficient to meet the ability of deoxyhemoglobin to accept H^+; this is referred to as the *isohydric carriage of CO_2* and is independent of hemoglobin's buffering ability

C. **Catalytic role of proteins: Enzymes**
   1. Many proteins function as biological catalysts; catalysts in living organisms are called enzymes
      a. Enzymes accelerate biochemical reaction rates by reducing the energy of activation needed to reach the transition state between reactant and product
      b. In enzyme-catalyzed biochemical reactions, the reactant molecule that binds first to the enzyme is called the substrate
      c. The **active site** of an enzyme are those amino acids that come into direct contact with the substrate and bind it
      d. Two different models describe how substrates bind to enzymes; the *lock-and-key model* assumes that the substrate and enzyme have complementary shapes that fit perfectly; the *induced fit model* assumes that the enzyme's active site changes its shape to fit the substrate's molecular configuration as the substrate binds
      e. Like all true catalysts, enzymes are not consumed by reactions they catalyze; after participating in a reaction, they recycle and catalyze additional reactions
      f. Enzymes are selective in the reactions they catalyze; many enzymes catalyze only one specific biological reaction
      g. Some enzymes require nonprotein groups called **cofactors** to function optimally; a cofactor may be a metal ion such as Zn^{2+}, or an organic molecule, such as NAD^+ (nicotinamide adenine dinucleotide); organic molecules that function as cofactors are called **coenzymes;** a cofactor bound to an enzyme is known as a *prosthetic group*
      h. An enzyme without its cofactor is called an **apoenzyme;** an enzyme associated with its cofactor is called a **holoenzyme**
      i. Enzymes requiring metal ions as cofactors are known as *metalloenzymes*
      j. Many organisms are unable to synthesize certain essential coenzymes and thus must consume these coenzymes in the diet; in humans, most the water-soluble vitamins are the dietary precursors of coenzymes
      k. **Isoenzymes** are either structurally different forms of the same enzyme or oligomeric proteins with various combinations of different subunits; isoenzymes usually occur in different tissues of the same organism, use different substrates, but catalyze similar types of reactions

l. Enzymes are classified as *allosteric* or *nonallosteric;* the activity of an allosteric enzyme is controlled by the binding of a second molecule, called an *effector* or *modulator;* the binding of the modulator may either decrease or increase the enzyme's activity; nonallosteric enzymes show no modulation through binding of other molecules
2. Enzymes require optimal conditions for optimal catalytic activity
    a. Because enzymes must bind precisely with their substrates, anything that changes the enzyme's three-dimensional shape changes its ability to catalyze reactions; enzymes that have been altered through physical or chemical means so that they are no longer active are called *denatured*
    b. Enzymes require an optimal temperature for optimal activity; enzymes are heat-labile and denature at temperatures exceeding normal physiologic temperature (37° C)
    c. Enzymes require optimal pH and ionic strength for optimal activity; the pH and ionic strength of the surrounding medium affect the charges on specific amino acid residues; the charges on the amino acid residues affect the molecule's configuration and therefore its activity
    d. Enzyme-catalyzed reactions require optimal concentrations of enzyme, substrate, and cofactors for the maximal reaction rate to be attained
3. Enzymes are named by attaching the suffix "ase" to the type of reaction they catalyze
    a. *Oxidoreductases* are class 1 enzymes; they catalyze oxidation-reduction reactions; for example, alcohol dehydrogenase catalyzes the oxidation of an alcohol to an aldehyde
    b. *Transferases* are class 2 enzymes; they catalyze the transfer of a group from one molecule to another; for example, phosphotransferase catalyzes the transfer of a phosphoryl group from one molecule to another
    c. *Hydrolases* are class 3 enzymes; they catalyze the breaking of covalent bonds using water; for example, peptidase hydrolyzes a peptide bond
    d. *Lyases* are class 4 enzymes; they either remove a group by splitting a bond and forming a double bond or add a group to a double bond and form a single bond; for example, decarboxylase removes a carboxyl group to form carbon dioxide
    e. *Isomerases* are class 5 enzymes; they catalyze internal atom rearrangements in a molecule; for example, racemase catalyzes the rearrangement (isomerization) of substituents on an α carbon atom
    f. *Ligases* are class 6 enzymes; they catalyze the formation of covalent bonds and therefore require energy, usually supplied by the hydrolysis of high-energy phosphate bonds; for example, pyruvate carboxylase catalyzes the formation of a carbon-to-carbon bond from pyruvate and carbon dioxide to form the compound oxaloacetate
4. Kinetics, the study of the rate (velocity) of reactions, uses the Michaelis-Menten equation to describe the rate of enzyme-catalyzed reactions; the Michaelis-Menten equation generally predicts the reaction rate for single substrate reactions involving nonallosteric enzymes only

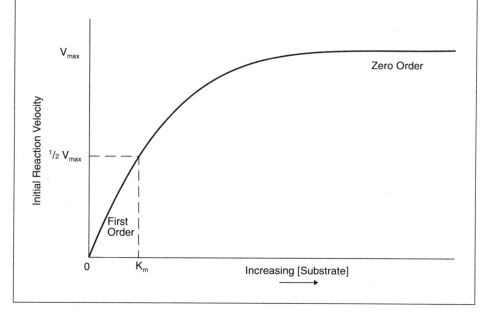

Initial Velocity of an Enzyme-Catalyzed Single-Substrate Reaction

For single-substrate reactions at a constant enzyme concentration, increasing the substrate concentration brings the reaction closer and closer to its maximum velocity (V_{max}); when the substrate concentration is low, the reaction velocity increases in direct proportion to the substrate concentration (displays first-order kinetics); when the substrate concentration is high enough to saturate all available enzyme active sites, the reaction velocity remains constant (displays zero-order kinetics).

a. In first-order reactions involving a single substrate, the initial reaction rate (V_i) depends only on the concentration of the substrate [S]; in zero-order reactions involving a single substrate, the rate of the reaction remains the same regardless of the concentration of the substrate because all active sites on the enzyme are already saturated with substrate

b. For single-substrate reactions at a constant enzyme concentration, nonallosteric enzymes display first-order kinetics as the concentration of substrate increases; when the substrate concentration is high enough to saturate all available enzyme active sites, the reaction displays zero-order kinetics (see *Initial Velocity of an Enzyme-Catalyzed Single-Substrate Reaction*)

c. The Michaelis-Menten equation describes enzyme-catalyzed reactions that follow the hyperbolic curve shown here (see *The Michaelis-Menten Equation,* page 62)

d. The concentration of a substrate required to achieve one-half of an enzyme's maximal velocity is a constant termed K_m; each enzyme has a unique, characteristic K_m; the lower the K_m, the higher the affinity of the enzyme for its substrate

> ## The Michaelis-Menten Equation
>
> The Michaelis-Menten equation describes the observed kinetics of a single substrate reaction catalyzed by a nonallosteric enzyme.
>
> $$V = \frac{V_{max}\,[S]}{[S] + K_m}$$
>
> V_{max} = maximum reaction rate for a given concentration of enzyme
> $[S]$ = substrate concentration
> K_m = Michaelis constant; the concentration of substrate required to achieve one-half the enzyme's maximal velocity.

    e. The Lineweaver-Burk plot is a plot of the reciprocal of the Michaelis-Menten equation; the reciprocal of the Michaelis-Menten equation is equivalent to a straight line equation of the form y = ax + b, where a is the slope of the line and b is the y intercept of the line (see *Lineweaver-Burk Plot of Enzyme Kinetic Data,* page 63)
    f. A Lineweaver-Burk plot provides a way to find the K_m and V_{max} for a particular reaction because its x-axis (1/[S]) intercept is $-1/K_m$ and its y-axis ($1/V_i$) intercept is $1/V_{max}$
    g. The Lineweaver-Burk plot also is called a double-reciprocal plot because 1/[S] and $1/V_i$ are plotted on the x-axis and y-axis respectively
    h. The slope of a Lineweaver-Burke plot is equal to K_m/V_{max}
5. The rate of enzyme-catalyzed reactions may be affected by the presence of inhibitors or by allosteric effects
    a. Inhibitors are substances that decrease the rate of an enzyme-catalyzed reaction
    b. *Competitive inhibitors* are substances with a molecular structure similar to the substrate; they compete with the substrate for binding at the enzyme's active site, but do not change the enzyme's affinity for the substrate; competitive inhibition may be overcome by increasing the substrate concentration
    c. *Noncompetitive inhibitors* are substances bearing no structural similarity to the substrate; they bind to the free enzyme or enzyme-substrate complex at a site other than the active site, thus reducing the enzyme's affinity for the substrate; noncompetitive inhibition cannot be overcome by increasing the substrate concentration
    d. *Irreversible inhibitors* form a covalent bond at the enzyme's active site, thus rendering the enzyme inactive
    e. A Lineweaver-Burk plot of reaction rates in the presence of an inhibitor can distinguish the type of inhibitor (see *Lineweaver-Burk Plots: Competitive and Noncompetitive Inhibitors,* page 64)
       (1) A competitive inhibitor alters the K_m but not the V_{max}; because the competitive inhibitor resembles and competes with the substrate for binding at the active site, more than the usual concentration of substrate (K_m) is needed to achieve one-half maximal velocity

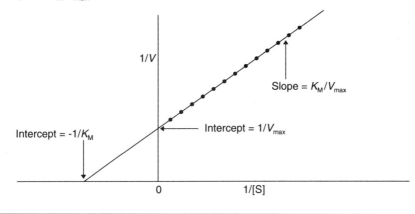

Lineweaver-Burk Plot of Enzyme Kinetic Data

The Lineweaver-Burk plot (also called a double-reciprocal plot) of kinetic data allows determination of K_m and V_{max} for a particular enzyme-catalyzed reaction; its x-axis (1/[S]) intercept is $-1/K_m$ and its y-axis intercept ($1/V_i$) is $1/V_{max}$; the slope of a Lineweaver-Burk plot is equal to K_m/V_{max}.

    (2) A noncompetitive inhibitor decreases the V_{max}, but does not alter the K_m; a noncompetitive inhibitor decreases the overall maximum reaction velocity (decreased V_{max}), but it does not change the concentration of substrate required to achieve one-half the lowered maximal velocity (K_m) because it does not bind at the active site

  f. Allosteric effects are changes in an enzyme's activity caused by conformational changes in the protein at sites distant from the active site

  g. Allosteric activators increase the reaction rate; allosteric inhibitors decrease the reaction rate

  h. Allosteric effects are important in ensuring that the body's metabolic processes are coordinated to meet the body's metabolic needs under different circumstances

6. The ***turnover number*** (k_{cat}) of an enzyme is a measure of an enzyme's activity; it is defined as the number of moles of substrate transformed per minute per mole of enzyme under optimal conditions

D. Structural role of proteins: Collagen

1. Collagen is an extracellular **fibrous** protein capable of providing great structural support; it is water-insoluble and the most abundant protein in vertebrates
2. The secondary structure of collagen is a triple helix (see Chapter 2, Protein Composition and Structure), a rigid structure that gives this protein high tensile strength
3. Collagen is composed of three individual polypeptide chains joined by interchain cross-links into a triple helix; collagen's secondary and tertiary structures include many intramolecular and intermolecular covalent cross-links
4. Humans have at least five types of collagen, each with a different amino acid composition

Lineweaver-Burk Plots: Competitive and Noncompetitive Inhibitors

As shown by the kinetic data plotted below, a competitive inhibitor (first plot) alters an enzyme's K_m but not its V_{max}; a noncompetitive inhibitor (second plot) decreases an enzyme's V_{max}, but does not alter its K_m.

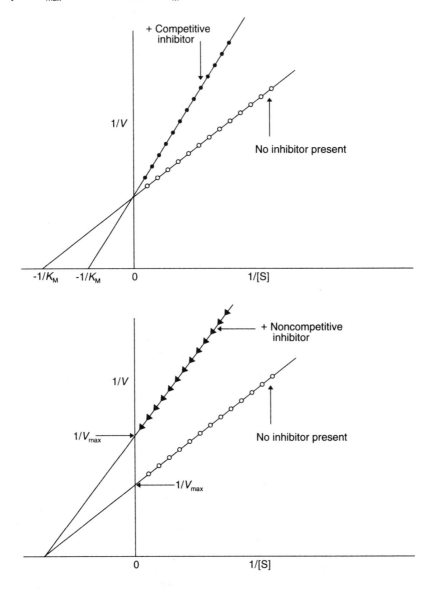

5. The body makes mature collagen from simpler, less mature forms of the protein called *procollagen* and *tropocollagen*

a. Procollagen, an immature and inactive form of collagen, is a **propeptide** (a peptide that must be modified before it is functional)
   (1) In the rough endoplasmic reticulum, procollagen is modified by hydroxylation of certain amino acid residues and addition of carbohydrate units; procollagen polypeptides containing these modifications form a triple-helical structure
   (2) The procollagen still contains extra N- and C-terminal amino acids, which must be removed
b. Hydrolysis of the peptide bonds linking the extra amino acids to procollagen yields tropocollagen, the next level of collagen maturity
c. Collagen, the mature protein, is produced when cross-linking occurs within and among the three helical chains to produce a triple helix
   (1) Intramolecular cross-links form between lysine residues
   (2) Intermolecular cross-links form between hydroxylysine and lysine residues

E. **Movement role of proteins: Actin and myosin**
  1. Actin and myosin, two proteins found in muscle cells, mediate muscle contraction
  2. Muscle cell cytoplasm (called sarcoplasm) contains fibers called *myofibrils*, which are composed of repeating functional units called *sarcomeres*
  3. Two types of protein filaments are found in the sarcomere: thick filaments composed primarily of myosin, and thin filaments composed primarily of actin but also containing the proteins tropomyosin and troponin
  4. Electron microscopy preparations of skeletal muscle reveal that a sarcomere contains alternating regions of a dark A band and a light I band
     a. The A band contains both thick and thin filaments
     b. The central, less dense, region of the A band, called the H zone, contains only thick filaments (myosin filaments)
     c. The H zone is bisected by a dark M-line, which results from the overlap of thick filaments in this region
     d. The I band, which contains only thin filaments (actin filaments), is bisected by a dense Z line, an area rich in proteins other than actin
  5. Sequential binding and release between the proteins of the thick and thin filaments is the biochemical event that brings about muscle movement
     a. Myosin, the major *structural protein* of the thick filaments, is a large, oligomeric, fibrous protein with globular portions that appear as projections or heads from the thick filament
     b. Myosin is also an *enzyme* that catalyzes the hydrolysis of adenosine triphosphate (ATP) to adenosine diphosphate (ADP); ATP hydrolysis provides the energy for muscle contraction
     c. Myosin is a *binding protein;* the binding and subsequent release of myosin and actin causes muscle movement
     d. The small, monomeric protein actin is the primary component of thin filaments; the proteins tropomyosin and troponin are bound to actin; tropomyosin is a double-stranded α helix; troponin is composed of three subunits, one of which binds calcium ions (Ca^{2+})

e. Calcium ions are the physiologic regulators of muscle contraction; Ca^{2+} is sequestered in the sarcoplasmic reticulum (analogous to the endoplasmic reticulum of nonmuscle cells) in resting skeletal muscle
      (1) A nerve impulse causes release of Ca^{2+} from the sarcoplasmic reticulum into the sarcoplasm
      (2) Released Ca^{2+} binds to troponin and induces a conformational change in both troponin and tropomyosin
   f. The sliding filament model explains how contraction occurs after muscles have been stimulated by Ca^{2+} movement into the sarcoplasm
      (1) The length of thick and thin filaments does not change; rather, the sarcomere shortens on contraction because the thick and thin filaments slide past each other
      (2) As calcium binds to troponin, the myosin globular head (called a cross bridge) binds to actin
      (3) The subsequent calcium-mediated conformational change in tropomyosin induces a tilting and sliding of the myosin globular head; sliding continues until the myosin globule detaches from actin and becomes available to bind to actin again at a new site

F. Storage role of proteins: Ferritin
1. Ferritin, a major protein involved in the storage of iron in body tissues, is an *apoprotein* (a protein without its prosthetic group) to which iron can reversibly bind
2. The structure of ferritin is well suited to its role
   a. Ferritin is a very large protein, containing 24 subunits, with a large internal cavity for iron storage
   b. Each molecule of ferritin can store as many as 4,500 iron ions, although in most individuals less than 3,000 are stored
   c. Channels from ferritin's surface to its cavity allow the uptake and release of iron
3. Ferritin facilitates the conversion of the ferrous—Fe(II) or Fe^{2+}—to the ferric—Fe(III) or Fe^{3+}—form of iron; iron is stored in its ferric form

G. Protective role of proteins: Immunoglobulins
1. Immunoglobulins (Ig), also called antibodies (Ab), are specific proteins produced in response to an antigen (Ag) (a specific foreign compound in the body)
   a. Immunoglobulins play a key role in a part of the body's immune system called the humoral immune system
   b. The humoral immune system relies on immunoglobulins to mark antigens for destruction by phagocytes (leukocytes, or specialized white blood cells that ingest and break down antigens as well as worn-out, diseased, or foreign cells)
   c. Immunoglobulins occur on the cell membrane of B-lymphocytes, where they act as binding sites for antigens
   d. Antigens also induce *clonal selection,* a proliferation of B cells and helper T cells (leukocytes that recognize and bind to the antigen)
   e. Binding of antigen to immunoglobulin at the surface of B-lymphocytes induces helper T-lymphocytes to bind to B-lymphocytes, which in turn causes B-lymphocytes to rapidly divide and multiply

f. Immature B-lymphocytes (called *plasma cells*) resulting from this rapid cell proliferation secrete large amounts of free immunoglobulin, which circulates in blood
g. This free immunoglobulin binds to free antigens and "marks" them for destruction by phagocytes

2. Immunoglobulins are glycoproteins (compounds in which protein is linked to carbohydrate), with protein constituting the major portion of the molecule
   a. The protein portion of a basic immunoglobulin molecule is a four-chain polypeptide, with each chain having both intra- and inter-chain disulfide bonds
   b. There are two identical *light* (L) *chains* and two identical *heavy* (H) *chains*, each with a *variable* and a *constant* region
      (1) The N-terminal portion of both heavy and light chains is the variable region because its amino acid sequence differs in each immunoglobulin
      (2) The C-terminal portions of the light chains and the remainders of the heavy chains are called the constant regions because their amino acid sequence is the same in each immunoglobulin
   c. The N-terminal region of each chain binds to antigen; the C-terminal region binds the chain to its host cell

3. Humans have five classes of immunoglobulins, each with a characteristic type of heavy chain (labeled α [alpha], δ [delta], ϵ [epsilon], γ [gamma], or μ [mu]); there are two different types of light chains, but these occur in all five classes of hemoglobin
   a. Immunoglobulin G (IgG), the most abundant immunoglobulin, accounts for approximately 70% of immunoglobulins produced; IgG is a monomer and is so named because it contains γ heavy chains
   b. IgM, usually the first immunoglobulin produced in response to an antigen, is a pentamer containing μ heavy chains
   c. IgA, the primary immunoglobulin found in such secretions as saliva, accounts for 15% to 20% of immunoglobulins; it is a monomer in humans and is so named because it contains α heavy chains
   d. IgD, a monomer present in only trace amounts, is of unknown function and is named for its δ heavy chains
   e. IgE is a monomer named for its ϵ heavy chains; it is associated with allergic reactions, such as hay fever, and is found most frequently in histamine-containing cells, such as mast cells and basophils

4. An immunoglobulin binds an antigen at the variable regions of its heavy and light chains
   a. The specific antigen-binding site is called the *hypervariable* region because its amino acid sequence differs in every immunoglobulin; the hypervariable region forms a cleft in which the antigen binds via many noncovalent bonds and interactions
   b. An antigen may be multivalent (contains more than one group that the body will recognize as foreign) and hence may elicit the production of more than one immunoglobulin
   c. Because each antigen induces production of a unique immunoglobulin, the immune system requires a very large number of immunoglobulins (1 million to 1 billion)

d. The production of this quantity of proteins is accomplished by rearrangements of the deoxyribonucleic acid (DNA) that encodes immunoglobulins (see Chapter 7, Transmission of Genetic Information); this biochemical feat allows great diversity in immunoglobulin production with minimal amounts of cell-stored DNA
5. Organisms exhibit two main classes of immune responses to invasion by foreign compounds
   a. In the *primary response* to antigen exposure, immunoglobulins (mostly IgM and IgG) appear in the blood approximately 7 days after invasion of an antigen; the blood concentration of immunoglobulins rises, reaches a plateau, then declines in approximately another 7 to 10 days
   b. Special B-lymphocytes called memory cells are already "primed" to react quickly against the antigen after the primary response occurs; they are part of the *secondary response* to an antigen
   c. Exposure to the same antigen for a second time provokes the memory cells to produce immunoglobulins more quickly, at higher concentrations, and for a longer duration than in the primary response

H. Role of proteins as hormones: Insulin
1. A *hormone* is a potent compound produced at one site in the body that travels to another target site to exert its effects
2. Hormones are chemically classified as either polypeptides, steroids, or amines; polypeptide hormones include growth hormone, adrenocorticotrophic hormone (ACTH), thyroid-stimulating hormone, oxytocin, and insulin
3. The hormone insulin is synthesized by the β cells of the pancreas and travels in the blood to its target sites (for example, to liver cells or adipose cells)
4. Insulin causes its target cells to take up glucose from the blood; the net effect is that the target cells have glucose to oxidize for energy and the blood glucose concentration drops
5. Two processing steps are required to produce insulin
   a. A compound called *preproinsulin* is synthesized on ribosomes in the pancreas; this compound contains an N-terminal **signal sequence** that targets preproinsulin to the membrane of the endoplasmic reticulum (see Chapter 7, Transmission of Genetic Information)
   b. The signal sequence is removed to yield *proinsulin,* which is targeted to the Golgi apparatus; in the Golgi apparatus, additional internal residues are removed to yield functional insulin
6. Mature insulin consists of two polypeptide chains covalently linked by the disulfide bonds of cystine residues

II. Function of Conjugated Proteins

A. General information
1. Many important biochemical proteins, called *conjugated* proteins, consist partly of protein and partly of some other molecule class
2. Proteins containing one or more covalently bound carbohydrate residue as prosthetic groups are *glycoproteins*
3. Proteins complexed with lipids in noncovalent interactions are *lipoproteins*

4. Proteins tightly bound to nucleic acids are *nucleoproteins*

B. **Glycoproteins**
   1. In glycoproteins, the carbohydrate, typically N-acetylglucosamine or N-acetylgalactosamine, is bonded to a serine, threonine, or asparagine residue in the process of ***glycosylation***
      a. Serine and threonine residues are linked to carbohydrates via the side-chain hydroxyl group oxygen; this process therefore is called O-glycosylation
      b. Asparagine residues are linked to carbohydrates via the side-chain amine group nitrogen; this process therefore is called N-glycosylation
   2. Carbohydrates are attached to proteins only after protein synthesis is completed; this process is called post-translational modification (See Chapter 7, Transmission of Genetic Information)
   3. The amount of carbohydrate bonded to protein varies from less than 5% to more than 80% carbohydrate
   4. Glycoproteins are typically found in cell membranes, where the carbohydrate portion is always located on the membrane's external surface
      a. The carbohydrates aid in orienting the protein into the correct position in the membrane's lipid bilayer
      b. Glycoproteins aid in transporting substances into and out of the cell
      c. Glycoproteins act as recognition and binding sites (receptors) for substances to be taken up by a cell
   5. Glycoproteins in erythrocyte membranes are antigens that determine whether an individual has type A (presence of N-acetylgalactosamine), B (presence of D-galactose), AB (presence of both N-acetylgalactosamine and D-galactose), or O (absence of both N-acetylgalactosamine and D-galactose) blood
   6. Glycoproteins are components of the mucus secreted by epithelial cells; mucus lubricates and protects tissues lined by these cells
   7. Many proteins found in blood plasma are glycoproteins, including immunoglobulins (previously described), clotting proteins, and several polypeptide hormones

C. **Lipoproteins**
   1. The lipid component of a lipoprotein may be a phospholipid, cholesterol, or triglyceride (see Chapter 5, Carbohydrate Structure, Function, and Metabolism); the amount of each lipid varies among the lipoproteins
   2. Lipoproteins are synthesized in the intestine and liver and then released into the blood
   3. Lipoproteins in blood function both in transport and metabolism of lipids; quantitative changes in blood lipoproteins are predictive of atherosclerosis
   4. The five major classes of lipoproteins are named based on their density, determined by ultracentrifugation (see Chapter 2, Protein Composition and Structure); the higher the protein content, the denser the lipoprotein
   5. Each type of lipoprotein has a distinct protein composition
      a. *Chylomicrons,* the least dense lipoproteins, appear in the blood only after eating; they are composed of approximately 2% protein and 98% lipid, the major portion of which are triglycerides

b. *Very-low-density lipoproteins* (VLDLs) transport triglycerides from the liver to other tissues; VLDLs are 5% to 10% protein and 90% to 95% lipid; triglycerides comprise about half the lipid content, with the other half divided between cholesterol and phospholipids
   c. *Intermediate-density lipoproteins* (IDLs) are 15% to 20% protein and 80% to 85% lipid; the lipid portion consists of approximately equal amounts of phospholipids, cholesterol, and triglycerides
   d. *Low-density lipoproteins* (LDLs) transport cholesterol from the liver, its site of synthesis, to other tissues; LDLs are 20% to 25% protein and 75% to 80% lipid; the lipid portion consists mostly of cholesterol, with smaller amounts of phospholipids and some triglycerides
   e. *High-density lipoproteins* (HDLs) transport cholesterol from tissues back to the liver for removal from the body; HDLs are 45% protein and 55% lipid; the lipid portion is mainly phospholipid
6. The protein portion of a lipoprotein, termed an *apolipoprotein,* is composed of a hydrophobic interior that associates with nonpolar lipids, and a hydrophilic exterior that associates with both polar lipids and the aqueous environment
   a. The apolipoprotein has a high degree of α-helical structure
   b. Every third or fourth amino acid residue has an ionic side chain, forming a polar edge along the helix; this polar region associates with the polar "head" portion of the phospholipid molecule and the aqueous extracellular fluid
   c. The amino acid residues containing hydrophobic side chains interact with the nonpolar phospholipid "tail" and other nonpolar lipids

D. Nucleoproteins
1. Nucleoproteins package the two nucleic acids, deoxyribonucleic acid (DNA) and ribonucleic acid (RNA)
2. In eukaryotic cells, DNA complexes with small basic proteins called **histones** to form chromosomes; half the chromosomal mass is DNA
3. Ribosomal RNA associates with proteins to form the ribonucleoprotein complexes or ribosomes of the cell; approximately two-thirds of the ribosomal mass is RNA

III. Protein Metabolism

A. General information
1. Proteins are composed of amino acids joined through peptide bonds; proteins ingested as part of the diet are broken down in the gut to their constituent amino acids, which are then absorbed from the small intestine directly into the blood
2. Cells within the body continually synthesize proteins (anabolism) and break them down (catabolism) to their component amino acids; this cyclic synthesis and breakdown of proteins helps cells regulate metabolism
3. To maintain the proper balance between protein degradation and synthesis, cells cleave proteins into their constituent amino acids as well as reuse, degrade, and synthesize amino acids as needed

4. Proteases are enzymes that cleave peptide bonds between specific amino acids, resulting in free amino acids; excess free amino acids (those not needed for protein synthesis) are neither stored nor excreted but are catabolized in the liver
   a. During amino acid catabolism, the α-amine group is removed and either recycled in the formation of new amino acids or excreted as urea
   b. Carbon backbones of amino acids are metabolized to pyruvate, to various intermediates of the tricarboxylic acid (TCA) cycle, or to acetyl coenzyme A (CoA) and thus become resources for gluconeogenesis or fatty acid synthesis

B. **Deamination and transamination of amino acids**
   1. The removal of the α-amine group from an amino acid is termed **deamination;** deamination occurs through two primary mechanisms: **transamination** and oxidative deamination (a process for deaminating glutamate only)
   2. Deamination of amino acids results in the formation of glutamate, aspartate, or ammonium ions (NH_4^+); high levels of NH_4^+ are toxic to humans
      a. Aquatic organisms can excrete NH_4^+ directly; birds and land-dwelling reptiles convert NH_4^+ to uric acid, which is excreted
      b. Land-dwelling vertebrate organisms, including humans, convert NH_4^+ to urea, which is excreted as a component of urine
   3. Transamination is the transfer of an α-amine group from an amino acid to an α-keto acid to yield a new amino acid and the α-keto acid of the original amino acid:

   Amino Acid$_a$ + α-keto acid$_b$ → α-keto acid$_a$ + amino acid$_b$

   a. Most transamination reactions use α-ketoglutarate or oxaloacetate as the amine group acceptor; the most common transamination reaction in human metabolism is the transfer of an amine group to α-ketoglutarate to form the amino acid glutamate

   Amine group from an amino acid + α-ketoglutarate →
   α-keto acid + glutamate

   b. Aminotransferases (previously called transaminases) are enzymes with specificity for transferring the amine group from particular amino acids to α-keto acids; for example, aspartate aminotransferase (AST) transfers the amine group from aspartate to α-ketoglutarate, forming glutamate and oxaloacetate; alanine aminotransferase (ALT) transfers the amine group from alanine to α-ketoglutarate, forming glutamate and pyruvate
   c. Pyridoxal phosphate (pyridoxal P), derived from pyridoxine (vitamin B_6), is a prosthetic group of aminotransferases
      (1) Pyridoxal P forms a covalent bond with the side chain amine group of a specific lysine residue at the active site of the aminotransferase; this bond is called a Schiff's base
      (2) Upon binding of the substrate to the aminotransferase, the α-amine group of the amino acid substrate displaces the amino group from the enzyme
   4. The other primary deamination reaction, oxidative deamination of glutamate is catalyzed by glutamate dehydrogenase, to produce NH_4^+ and α-ketoglutarate

a. Unlike other dehydrogenases, glutamate dehydrogenase uses either NAD^+ or nicotinamide adenine dinucleotide phosphate ($NADP^+$); glutamate dehydrogenase is the only known enzyme that can use both NAD^+ or $NADP^+$

b. Guanosine triphosphate (GTP) and ATP inhibit glutamate dehydrogenase, while guanosine diphosphate (GDP) and ADP activate it; a decrease in the **energy charge** accelerates the oxidation of amino acids for greater energy production

C. The urea cycle

1. Excretion of NH_4^+ as urea occurs in a series of reactions called the urea cycle; urea is synthesized in the liver, transported in the blood, and removed from the blood by the kidneys (see *The Urea Cycle*)
2. The urea cycle consumes four high-energy bonds in three molecules of ATP for every two amine groups eliminated through urea formation; the first two steps of the cycle take place in the mitochondrion; the remainder of the cycle takes place in the cytosol; the overall reaction for the production of urea through the urea cycle is as follows:

$$NH_3 + HCO_3^- + \text{Aspartate} + 3\,ATP \rightarrow$$
$$\text{Urea} + \text{Fumarate} + 2ADP + 2P_i + AMP + PP_i$$

  a. In *Step 1*, carbamoyl phosphate is synthesized from NH_4^+, CO_2, H_2O, and two ATP molecules (catalyzed by carbamoyl phosphate synthetase); each ATP molecule is hydrolyzed to ATP, consuming two high-energy bonds

  b. In *Step 2*, carbamoyl phosphate combines with ornithine to form citrulline; the enzyme ornithine transcarbamoylase catalyzes the reaction; citrulline diffuses from the mitochondria and into the cytosol; ornithine is similar to oxaloacetate of the TCA cycle (see Chapter 3, Bioenergetics) in that it is regenerated with each turn of the cycle

  c. In *Step 3*, the α-amine group of aspartate provides the second amine group that ultimately becomes urea; the enzyme argininosuccinate synthetase catalyzes the formation of argininosuccinate; this reaction is driven by the hydrolysis of the last molecule of ATP in this cycle to AMP, thus consuming two more high-energy bonds

  d. In *Step 4*, the enzyme argininosuccinase catalyzes the cleavage of argininosuccinate into arginine and fumarate

  e. In *Step 5*, the enzyme arginase cleaves urea from arginine and ornithine is regenerated for another turn of the urea cycle

D. Catabolism of amino acid carbon

1. The carbon skeletons of deaminated amino acids are converted to compounds that can produce glucose, be oxidized in the TCA cycle, or generate ketone bodies, fatty acids, or hormones; the results depend on the structure of the initial amino acid
2. All carbon skeletons from the twenty amino acids can be converted into the following seven molecules: acetyl CoA, acetoacetyl CoA, pyruvate, or the TCA cycle intermediates α-ketoglutarate, succinyl CoA, fumarate, or oxaloacetate

The Urea Cycle

The first two steps of the urea cycle take place in the mitochondrion; the last three steps take place in the cytosol. Circles on the reaction diagram below indicate where reactants move from mitochondrion to cytosol and vice versa.

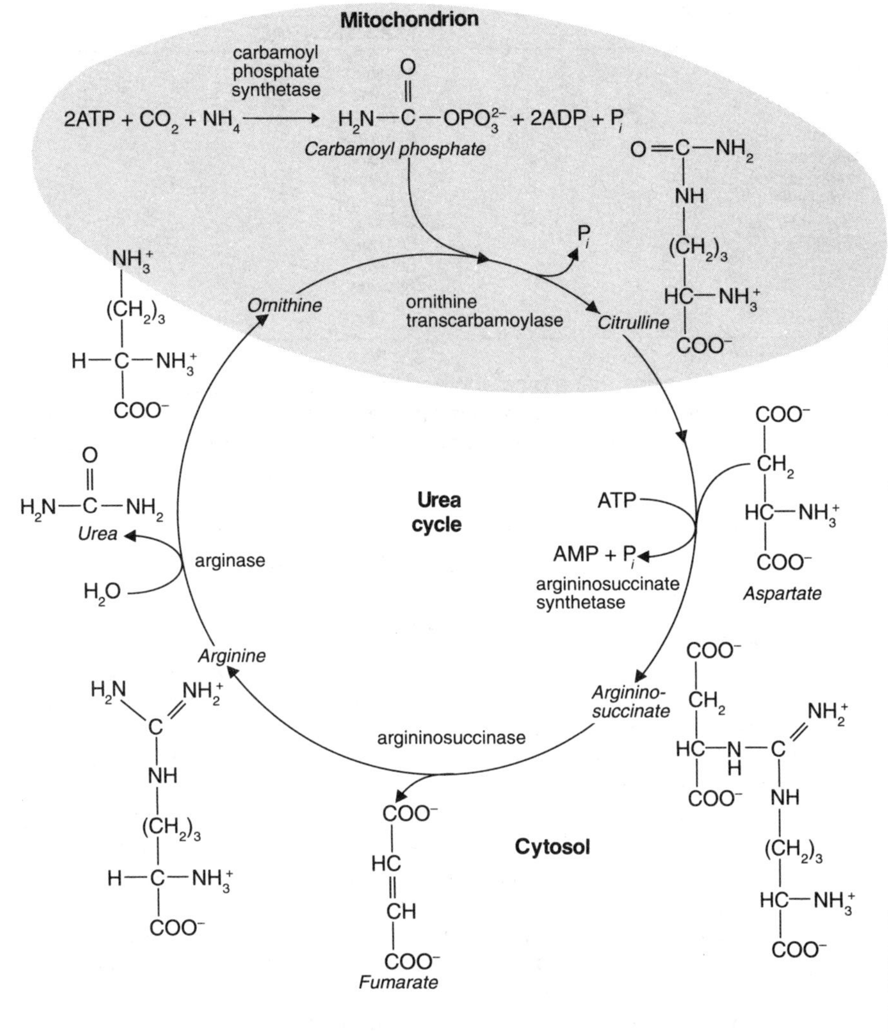

Mitochondrion

$$2ATP + CO_2 + NH_4 \xrightarrow{\text{carbamoyl phosphate synthetase}} H_2N-C(=O)-OPO_3^{2-} + 2ADP + P_i$$

Carbamoyl phosphate

Ornithine → ornithine transcarbamoylase → Citrulline

Citrulline structure: $O=C-NH_2$, NH, $(CH_2)_3$, $HC-NH_3^+$, COO^-

Ornithine structure: NH_3^+, $(CH_2)_3$, $H-C-NH_3^+$, COO^-

Urea: $H_2N-C(=O)-NH_2$

arginase, H_2O

Urea cycle

Arginine: H_2N, NH_2^+, C, NH, $(CH_2)_3$, $H-C-NH_3^+$, COO^-

argininosuccinase

Fumarate: COO^-, HC, $\parallel$, CH, COO^-

ATP → AMP + P_i, argininosuccinate synthetase

Aspartate: COO^-, CH_2, $HC-NH_3^+$, COO^-

Argininosuccinate: COO^-, CH_2, $HC-N(H)-C(=NH_2^+)-NH-(CH_2)_3-HC(NH_3^+)-COO^-$

Cytosol

a. Amino acids that are catabolized to acetyl CoA or acetoacetyl CoA are called **ketogenic** amino acids because excess acetyl CoA or acetoacetyl CoA leads to the formation of ketone bodies

> ### Essential and Nonessential Amino Acids for Humans
>
> Essential amino acids are those that are necessary for human life but cannot yet be synthesized by the body. Humans obtain essential amino acids from the diet. Different organisms have different essential amino acids, depending on which enzymes are lacking for synthesis of a particular amino acid.
>
> | ESSENTIAL | NONESSENTIAL |
> |---|---|
> | Histidine | Alanine |
> | Isoleucine | Arginine |
> | Leucine | Asparagine |
> | Lysine | Aspartate |
> | Methionine | Cysteine |
> | Phenylalanine | Glutamate |
> | Threonine | Glutamine |
> | Tryptophan | Glycine |
> | Valine | Proline |
> | | Serine |
> | | Tyrosine |

   b. Amino acids that are catabolized to pyruvate, α-ketoglutarate, succinyl CoA, fumarate, or oxaloacetate are called **glucogenic;** they can be converted to phosphoenolpyruvate and then to glucose via gluconeogenesis; glucose cannot be synthesized from acetyl CoA or acetoacetyl CoA
   c. Some amino acids are both ketogenic and glucogenic because their catabolism proceeds through steps that yield both ketogenic and glucogenic end products

E. **Amino acid anabolism**
   1. After reduction by certain bacteria that dwell on the roots of leguminous plants, nitrogen (N_2) from the atmosphere is incorporated into amino acids as NH_4^+; this process is called *nitrogen fixation*
   2. In eukaryotes inorganic nitrogen, as NH_4^+, is incorporated into amino acids as glutamate and glutamine
      a. NH_4^+ reacts with α-ketoglutarate (produced in the TCA cycle), forming glutamate; glutamate dehydrogenase catalyzes this reaction
      b. Glutamate reacts with NH_4^+, to form glutamine and H^+; glutamine synthetase catalyzes the reaction
      c. In contrast, prokaryotes use glutamine as the nitrogen donor to produce glutamate by the reductive amination of α-ketoglutarate; glutamate synthase catalyzes this reaction
   3. The eleven amino acids that humans can synthesize are called *nonessential;* the term means that these amino acids can be synthesized by the body and therefore do not have to be obtained in the diet; it does not mean that they are unnecessary for life (see *Essential and Nonessential Amino Acids for Humans*)
   4. Humans must obtain nine amino acids from the diet because they lack the enzymes necessary to synthesize these molecules; these amino acids are called *essential*

5. The carbon skeletons of all the nonessential amino acids are derived from the intermediates of glycolysis, the pentose phosphate pathway, the TCA cycle, and other amino acids
   a. Alanine is produced from pyruvate and glutamate, catalyzed by alanine aminotransferase
   b. Aspartate is produced from oxaloacetate and glutamate, catalyzed by aspartate aminotransferase
   c. Asparagine is synthesized from aspartate and glutamine; this reaction requires the hydrolysis of ATP to AMP and pyrophosphate
   d. Tyrosine is produced by the hydroxylation of phenylalanine (an essential amino acid) in a reaction catalyzed by phenylalanine hydroxylase and requiring tetrahydrobiopterin as a cofactor
   e. Glutamate is synthesized from NH_4^+ and α-ketoglutarate; the reaction, catalyzed by glutamate dehydrogenase, requires either NADPH or NADH as a cofactor
   f. Glutamine, proline, and arginine are produced from glutamate
   g. Serine is synthesized from 3-phosphoglycerate, an intermediate of glycolysis
   h. Glycine is synthesized from serine
   i. Cysteine is synthesized from serine and methionine (an essential amino acid); *S*-adenosylmethionine (SAM), an "active" form of methionine, is a methyl group donor

Study Activities

1. Describe the hemoglobin molecule in terms of its structure, O_2- and BPG-binding sites, and conformation in the oxygenated and deoxygenated states.
2. Describe the enzyme binding sites of the following compounds: substrate, competitive inhibitor, noncompetitive inhibitor, irreversible inhibitor, prosthetic group, allosteric inhibitor.
3. Discuss the information that can be obtained from a Lineweaver-Burk plot of enzyme kinetic data.
4. Compare protein processing of collagen and insulin.
5. Use the sliding filament model of muscle contraction to discuss how proteins function in movement.
6. Compare and contrast the five classes of immunoglobulins in terms of composition, function, and presence in the body.
7. Describe functions of glycoproteins, lipoproteins, and nucleoproteins.
8. Outline the reactions of the urea cycle and state the function of this cycle.
9. Differentiate transamination from deamination; give specific examples of each.
10. Define the terms listed in bold-faced type throughout this chapter.

5
Carbohydrate Structure, Function, and Metabolism

Objectives

After studying this chapter, the reader should be able to:
- Classify carbohydrates according to their number of carbon atoms, functional groups, and ring forms.
- Classify carbohydrates as polymers of individual sugar molecules.
- Describe the role of glycogen in maintaining blood glucose levels; discuss its synthesis, degradation, and metabolic regulation.
- Compare and contrast glycolysis with gluconeogenesis.
- Discuss the purpose and the reactions of the pentose phosphate pathway.

I. Carbohydrate Structure and Function

A. General information
1. Carbohydrates (commonly called sugars) are aldehydes or ketones containing multiple hydroxyl groups; the word "carbohydrate" simply indicates a hydrate of a carbon-containing compound
2. Carbohydrates exist as individual molecules, called *monosaccharides,* or as a number of individual carbohydrate molecules joined by glycosidic bonds; carbohydrates joined through covalent bonds are called *oligo-* (a few) or *poly-* (more than three) *saccharides,* depending on how many carbohydrates are joined as one molecule
3. Starch is the storage form of carbohydrates in plants; glycogen is the storage form of carbohydrates in animals; both are polysaccharides
4. The monosaccharides ribose and deoxyribose comprise part of ribonucleic acid (RNA) and deoxyribonucleic acid (DNA), respectively; these molecules are responsible for transmitting genetic information (see Chapter 7, Transmission of Genetic Information)
5. Other biochemically important carbohydrates include glucose (a monosaccharide that is the principle source of energy for most cells), fructose (a monosaccharide found in fruit and honey), and lactose (a disaccharide found in milk)
6. Polysaccharides such as cellulose provide structure in the cell walls of bacteria and plants

7. Carbohydrates may be complexed with proteins to form glycoproteins (such as the antibodies of the immune system) and with lipids to form glycolipids (such as the cerebrosides characteristic of neuronal cell membranes)

B. **Levels of carbohydrate complexity**
   1. **Monosaccharides** are simple sugars (that is, not joined with any other sugars) having three or more carbon atoms with the general formula $(CH_2O)_n$ where n represents the number of carbon atoms in the sugar
      a. Although carbohydrates occur as both D and L isomeric forms (see Chapter 1, Basics of Biochemistry), almost all naturally occurring sugars are the D isomer
      b. Monosaccharides are classified as aldoses or ketoses, depending upon whether they contain an aldehyde or a ketone group
      c. The simplest monosaccharides are the trioses (three-carbon sugars); glyceraldehyde is the simplest three-carbon aldose and dihydroxyacetone is the simplest three-carbon ketose

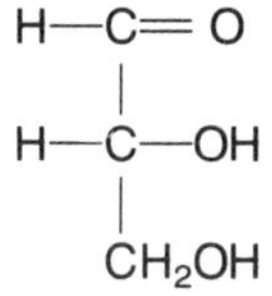

$$\begin{array}{cc}
\text{H—C}=\text{O} & \text{CH}_2\text{OH} \\
| & | \\
\text{H—C—OH} & \text{C}=\text{O} \\
| & | \\
\text{CH}_2\text{OH} & \text{CH}_2\text{OH} \\
\text{D-Glyceraldehyde} & \text{Dihydroxyacetone}
\end{array}$$

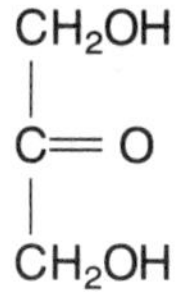

      d. Ribose, an important pentose (five-carbon sugar), is a constituent of RNA (see Chapter 7, Transmission of Genetic Information) and also an intermediate in a metabolic pathway called the pentose phosphate pathway

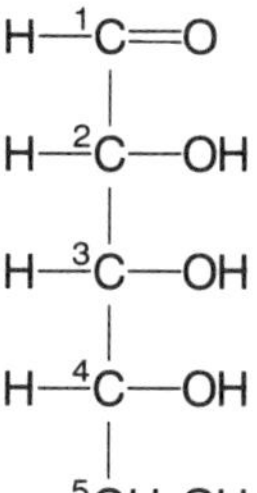

$$\begin{array}{c}
\text{H—}^1\text{C}=\text{O} \\
| \\
\text{H—}^2\text{C—OH} \\
| \\
\text{H—}^3\text{C—OH} \\
| \\
\text{H—}^4\text{C—OH} \\
| \\
^5\text{CH}_2\text{OH} \\
\text{D-Ribose}
\end{array}$$

      e. Two important metabolic hexoses (six-carbon sugars) are glucose, an aldose, and fructose, a ketose (see *Glucose and Fructose in Fischer Projection Formulas,* page 79)
      f. In solution, pentoses and hexoses are found almost exclusively in ring forms; Haworth projections are representations of monosaccharide ring structures

(1) The designations α and β indicate whether the hydroxyl group of the anomeric carbon (the new asymmetric center of a carbohydrate in a ring form) lies below (α) or above (β) the plane of the ring when the anomeric carbon is to the right, and the ring is numbered in a clockwise direction in a Haworth projection (in the illustrations on these pages, asterisks indicate the anomeric carbon)

(2) In ribose, the carbonyl carbon (C-1) can react with the C-4 hydroxyl group, with rotation around the C-3 to C-4 bond, to form a five-membered ring called a *furanose;* the C-1 of ribose also can react with the C-5 hydroxyl group, to form a six-membered ring called a *pyranose*

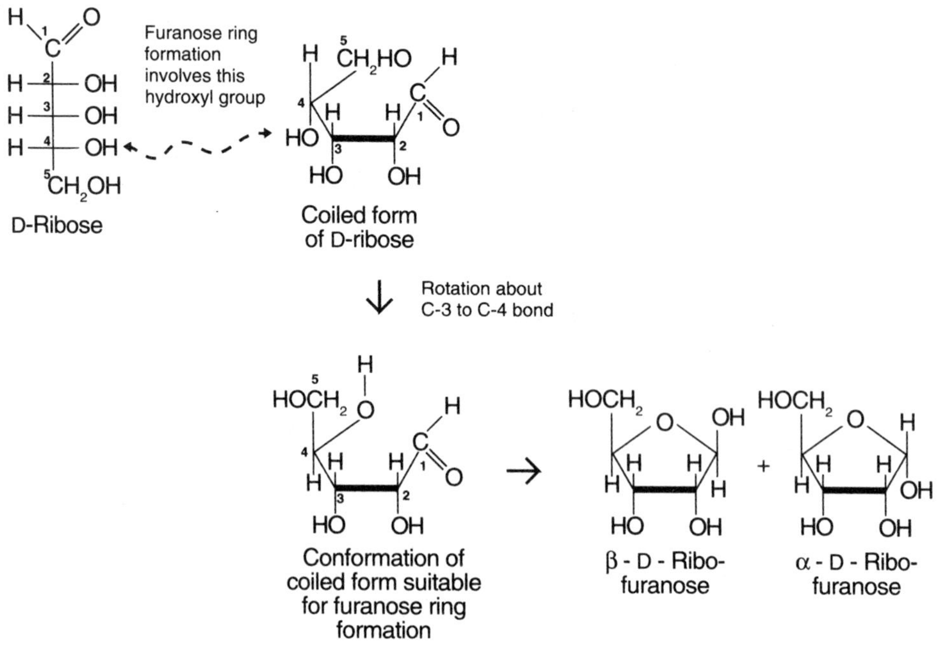

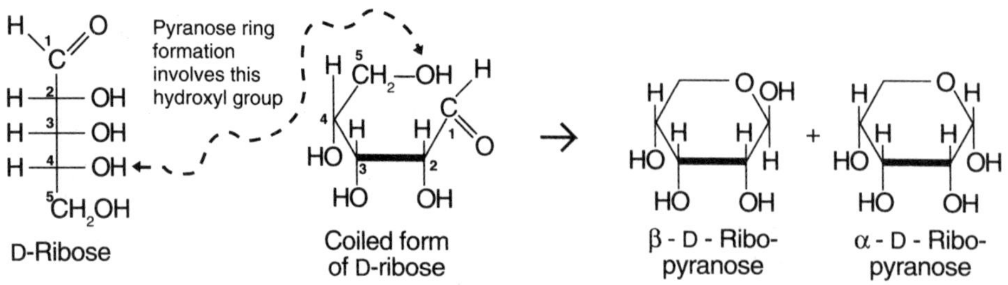

Glucose and Fructose in Fischer Projection Formulas

The Fischer projection formulas of these two important monosaccharides reveal the carbon skeleton of the sugars, but not their true secondary structures; monosaccharides exist primarily in ring form.

D-Glucose

$H-^1C=O$
$H-^2C-OH$
$HO-^3C-H$
$H-^4C-OH$
$H-^5C-OH$
$^6CH_2OH$

D-Fructose

CH_2OH
$C=O$
$HO-C-H$
$H-C-OH$
$H-C-OH$
CH_2OH

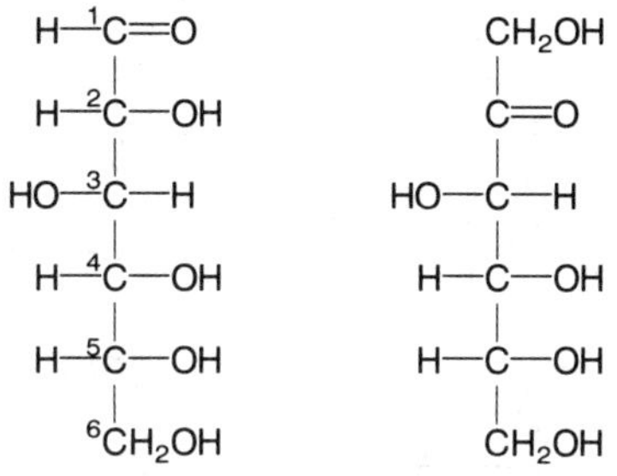

(3) In glucose solutions, the carbonyl carbon (C-1) reacts with the C-5 hydroxyl group to form a pyranose ring

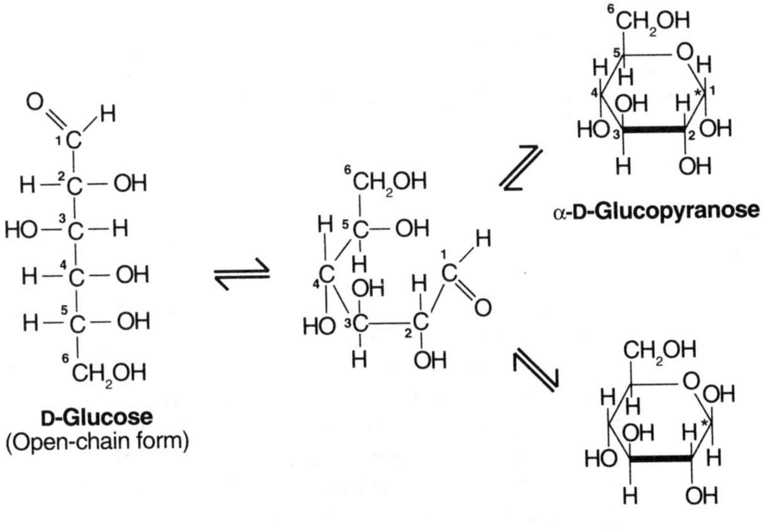

D-Glucose
(Open-chain form)

α-D-Glucopyranose

β-D-Glucopyranose

Carbohydrate Structure, Function, and Metabolism

Structures of Sucrose and Lactose

Two common disaccharides, sucrose and lactose, demonstrate two different glycosidic bonds; sucrose has a glycosidic bond between C-1 of glucose and C-2 of fructose; lactose has a glycosidic bond between C-1 of galactose and C-4 of glucose.

Sucrose

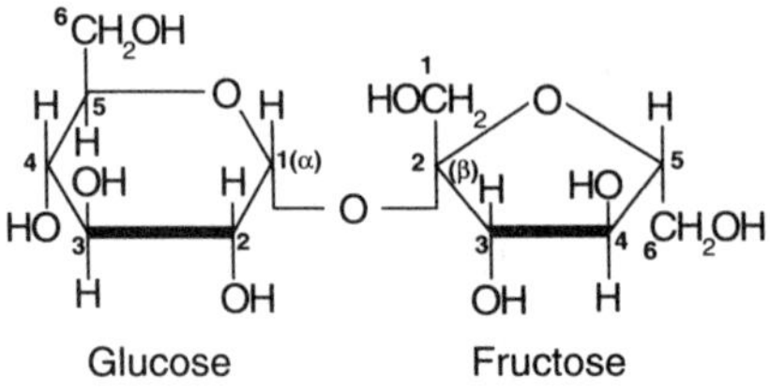

Glucose — Fructose

Lactose

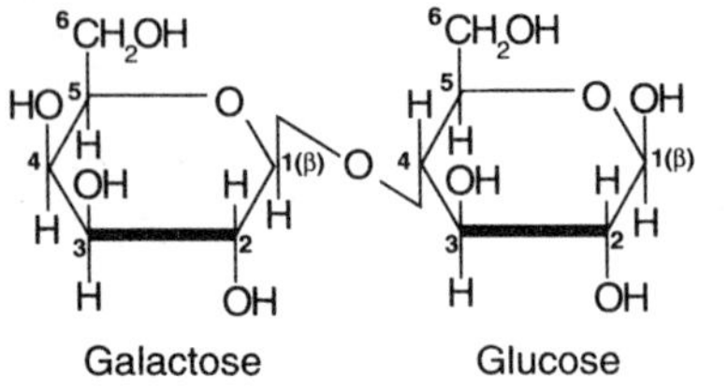

Galactose — Glucose

---

(4) In fructose solutions, the ketone carbon at C-2 reacts with the C-5 hydroxyl group to form a furanose ring

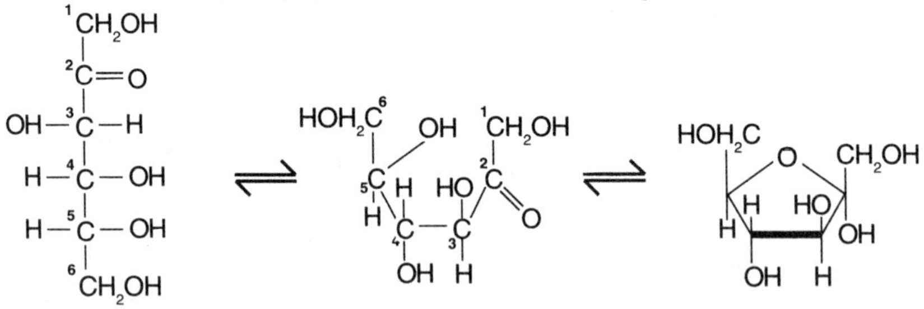

D-Fructose ⇌ ⇌ **α-D-Fructofuranose**
(A ring form of fructose)

  g. Monosaccharides are typically phosphorylated, a modification that imparts certain characteristics to a sugar
    (1) Phosphorylated carbohydrates are sources of energy because the phosphate bond releases energy when it is hydrolyzed
    (2) Phosphorylated carbohydrates are anionic; their negative charge renders them able to participate in electrostatic interactions
    (3) Phosphorylated carbohydrates are reactive intermediates in many metabolic pathways and in the synthesis of purines and pyrimidines (nitrogenous bases found in DNA and RNA)
2. A *disaccharide* is composed of two monosaccharides joined together by an O-glycosidic bond between the hydroxyl group at the anomeric carbon of the first carbohydrate and a carbon of the second carbohydrate; two of the most common disaccharides are sucrose and lactose (see *Structures of Sucrose and Lactose*)
  a. Sucrose (glucose-α,(1→2)-β-fructose) contains glucose and fructose linked between C-1 of α-D glucose and C-2 of β-D fructose, with the bond from glucose projecting below (α) and the bond from fructose projecting above (β) the plane of the rings

b. Lactose is galactose-$\beta(1\rightarrow 4)$-glucose
3. An **oligosaccharide** is a general term used for a carbohydrate containing between two and eight monosaccharides; alternatively, a prefix indicating the number of carbon atoms may be used to denote the specific number of monosaccharides (for example, a disaccharide contains two monosaccharides)
4. **Polysaccharides,** also called complex carbohydrates, contain more than eight of the same or different monosaccharides

C. **Storage forms of glucose**
   1. Glycogen, the main storage form of glucose in animal cells, is a large branched polymer of glucose units linked by $\alpha(1\rightarrow 4)$ and $\alpha(1\rightarrow 6)$ glycosidic bonds
      a. Branching, which occurs approximately once per ten glucose units, increases the volubility of glycogen and creates additional terminal units, points at which glycogen synthesis and degradation occur
      b. Glycogen is stored as granules in the cytosol, primarily in liver and skeletal muscle cells
      c. Glycogen catabolism and synthesis (a primary function of the liver) help regulate the concentration of glucose in the blood
         (1) When the blood glucose concentration is low, glycogen is catabolized to glucose in a process called **glycogenolysis** and then released into the blood
         (2) When the blood glucose concentration is high (implying an adequate glucose supply), glucose is removed from the blood for glycogen synthesis; this process, called **glycogenesis,** permits excess glucose to be stored
   2. Starch and cellulose are the main storage forms of glucose in plant cells
      a. Starch, as a glucose reservoir for metabolic reactions, occurs in the forms of amylose and amylopectin
         (1) Amylose is composed of glucose units linked by $\alpha(1\rightarrow 4)$ glycosidic bonds, forming only straight chain structures
         (2) Amylopectin is composed of glucose units linked by $\alpha(1\rightarrow 4)$ and $\alpha(1\rightarrow 6)$ glycosidic bonds, forming straight chain and branched structures, respectively; branching occurs less frequently than in glycogen, approximately once per thirty glucose units
         (3) Humans degrade amylose and amylopectin by the action of the enzyme α-amylase, found in the salivary glands and pancreas
      b. Cellulose serves a structural role in plants; it is a straight chain of glucose units linked by $\beta(1\rightarrow 4)$ glycosidic bonds
         (1) The $\beta(1\rightarrow 4)$ linkages of cellulose impart the high tensile strength characteristic of woody and fibrous plants
         (2) Humans lack the enzymes (cellulases) necessary to digest cellulose, resulting in their inability to digest certain parts of plants

II. Carbohydrate Metabolism

A. **General information**
   1. Carbohydrates serve as sources of energy, storage forms of energy, intermediates in metabolic processes, and structural compounds

2. Complex carbohydrates are catabolized to glucose, which in turn is used to generate energy as adenosine triphosphate (ATP) and reducing power in the form of reduced nicotinamide adenine dinucleotide phosphate (NADPH) (see Chapter 3, Bioenergetics)
3. Glucose is stored as glycogen when the glucose supply is plentiful; conversely, glycogen is used to produce glucose when the glucose supply is inadequate
4. When the need for glucose exceeds its supply, certain specialized tissues can convert noncarbohydrates to glucose
    a. In mammals, the liver and kidneys are the principle organs for **gluconeogenesis,** synthesis of new glucose molecules from noncarbohydrate precursors
    b. Humans can synthesize new glucose from pyruvate, lactate, tricarboxylic acid (TCA) cycle intermediates, glycerol, and most, but not all, amino acids

B. Glycogen degradation (glycogenolysis)
1. When the blood glucose concentration is low, glycogen is degraded and transported by the blood to organs that must have a constant glucose supply, such as the brain (which lacks energy stores) and skeletal muscles (which have large stores of glycogen but utilizes large quantities of glucose)
2. Recall that glycogen is a large branched polymer of glucose units, linked by $\alpha(1\rightarrow4)$ and $\alpha(1\rightarrow6)$ glycosidic bonds that form straight chains and branches, respectively; the highly branched structure makes glycogen a starch with many nonreducing ends (final glucose residues without a free anomeric carbon)
3. In the first step of glycogen degradation, the enzyme *glycogen phosphorylase* cleaves glucose one residue at a time, beginning at a nonreducing end of a glycogen branch
    a. The enzyme cleaves at the $\alpha(1\rightarrow4)$ glycosidic bond between the final residue and the next sugar, and requires pyridoxal 5'-phosphate (a derivative of vitamin B_6) as a cofactor
    b. During the cleavage, the enzyme also catalyzes the phosphorolysis (addition of inorganic phosphate) of the glucose residue to form glucose 1-P; the glycogen remains in its original state but with one fewer glucose unit
    c. Glycogen phosphorylase cleaves glucose from glycogen only as a result of a cascade of events that convert the enzyme from its inactive to its active form
        (1) Glycogen phosphorylase exists in a *b* form (dephosphorylated) and an *a* form (phosphorylated); only the *a* form is the active enzyme
        (2) Glycogen phosphorylase *b* is converted to glycogen phosphorylase *a* through the action of another enzyme, *phosphorylase kinase*
        (3) To convert inactive glycogen phosphorylase *b* to active glycogen phosphorylase *a*, phosphorylase kinase simply adds inorganic phosphate to the enzyme
        (4) Glycogen phosphorylase *a* is converted to glycogen phosphorylase *b* through the action of the enzyme phosphoprotein phosphatase

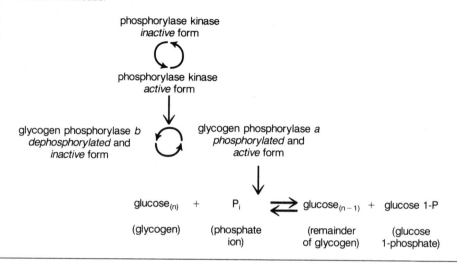

    (5) Phosphorylase kinase, like glycogen phosphorylase, also exists in an active *a* and an inactive *b* form; it is activated by a cyclic adenosine monophosphate (cAMP)-dependent protein kinase; it is also partially activated by calcium ions; one of its subunits is the calcium binding protein calmodulin (see *Enzyme Control of Glycogen Degradation [Glycogenolysis]*)

  d. The glycogen phosphorylase reaction removes glucose units one at a time until four units remain on a branch; the limit branch is a four-unit branch

4. When four glucose units remain on a branch, a *debranching enzyme* moves the three outer units of the limit branch to the nonreducing end of another branch to be available for removal by glycogen phosphorylase
5. The last glucose unit of the original branch is attached to glycogen through an α(1→6) glycosidic bond; the debranching enzyme hydrolyzes this final glycosidic bond
6. The enzyme *phosphoglucomutase* catalyzes the conversion of glucose 1-P to glucose 6-P
  a. There is no glucose 6-P transport system across cell membranes; glucose 6-P remains in the cell where it was created
  b. In muscle cells, glucose 6-P can enter glycolysis
  c. Liver, intestine, and kidney cells contain the enzyme glucose 6-phosphatase, which removes the phosphate group from glucose 6-P to form glucose; glucose then can enter the blood for delivery to cells that need it
    (1) Brain and skeletal muscle lack glucose 6-phosphatase; the absence of this enzyme traps glucose 6-P in these cells that need it quickly

Glycogen Synthesis

Glucose, as glucose 1-P, is activated by complexing with uridine diphosphate (UDP) to form UDP-glucose before its incorporation into glycogen.

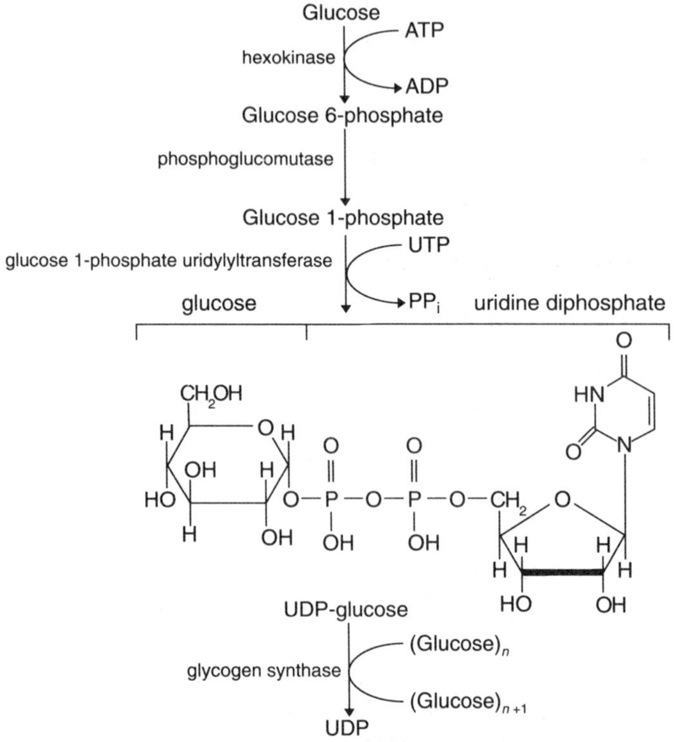

Glucose
↓ hexokinase (ATP → ADP)
Glucose 6-phosphate
↓ phosphoglucomutase
Glucose 1-phosphate
↓ glucose 1-phosphate uridylyltransferase (UTP → PP$_i$)
glucose + uridine diphosphate

UDP-glucose
↓ glycogen synthase ((Glucose)$_n$ → (Glucose)$_{n+1}$)
UDP

(2) Muscles store glycogen to have a ready supply of glucose; brain cells cannot live for more than 5 minutes without glucose; lack of glucose 6-phosphatase in these cells ensures that their intracellular glucose supply is not sent to less glucose-dependent cells

C. Glycogen synthesis (glycogenesis)
  1. Glycogen synthesis occurs when blood glucose levels are high, allowing excess glucose to be stored
     a. The liver is a major glycogen storage site; liver glycogen regulates glucose levels in the blood
     b. Muscle is a major glycogen storage site; muscle glycogen is a fuel reservoir for ATP synthesis, required by active muscle
  2. In the first step of glycogen synthesis, the enzyme hexokinase catalyzes the conversion of glucose to glucose 6-P, just as in glycolysis (see *Glycogen Synthesis*)
  3. The enzyme phosphoglucomutase catalyzes the conversion of glucose 6-P to glucose 1-P

4. Glucose 1-P is activated by complexing with uridine triphosphate (UTP) to form uridine diphosphate (UDP)-glucose
   a. Glucose 1-P reacts with UTP, catalyzed by glucose 1-phosphate uridylyltransferase (also called UDP-glucose pyrophosphorylase), to form UDP-glucose (UDPG) and pyrophosphate
   b. Pyrophosphate is rapidly hydrolyzed to two molecules of inorganic phosphate in a reaction catalyzed by pyrophosphatase
      (1) Hydrolysis of pyrophosphate has a large negative free energy, but the formation of UDPG has a free energy change of nearly zero
      (2) The free energy release that accompanies the hydrolysis of pyrophosphate drives the reaction that forms UDPG
      (3) The formation of UDPG is an essentially irreversible reaction
5. UDPG is added at the C-4 hydroxyl group of a nonreducing end of the glycogen molecule
   a. *Glycogen synthase,* which catalyzes this reaction, adds glucose units only if the glycogen molecule already contains more than four glucose units
   b. Glycogen synthase has an active (dephosphorylated) form and an inactive (phosphorylated form); each of these forms is regulated by a separate enzyme
      (1) Phosphoprotein phosphatase dephosphorylates the glycogen synthase to its active form
      (2) The enzyme cAMP-dependent protein kinase phosphorylates the glycogen synthase to its inactive form
6. Branches—α(1→6) linkages—are added to the growing glycogen chain through the action of α(1→4)-glucan-branching enzyme, which transfers several glucose units to an interior glucose unit

D. Regulation of glycogen degradation and synthesis
1. *Covalent modification* of individual enzymes controls their activities
   a. Phosphorylation of glycogen synthase, the enzyme catalyzing glycogen synthesis, has the opposite effect of phosphorylation of glycogen phosphorylase, the enzyme catalyzing glycogen degradation
   b. When phosphorylated, glycogen phosphorylase is active and glycogen synthase is inactive; dephosphorylated glycogen phosphorylase is inactive while dephosphorylated glycogen synthase is active and vice versa
2. The *hormones* insulin, glucagon, and epinephrine control whether glycogen synthesis or degradation occurs
   a. Insulin lowers the blood glucose concentration by increasing the uptake of glucose by the liver and other cells, thereby promoting glycogen synthesis from the excess blood glucose; glucose inhibits glycogen phosphorylase *a* (the active enzyme), to prevent the degradation of glycogen
   b. The effect of glucagon and epinephrine is the reverse of insulin; glucagon and epinephrine increase the blood glucose level by stimulating glycogen breakdown
      (1) Epinephrine causes glycogen degradation, primarily in skeletal muscles; glucagon causes glycogen degradation, primarily in the liver

(2) The effects of both hormones are mediated by the cAMP
(3) Epinephrine and glucagon bind to **receptors** in the plasma membrane of skeletal muscle and liver cells, respectively; their binding causes a signal coupling protein to activate adenylate cyclase, the enzyme that catalyzes the formation of cAMP
(4) The cAMP activates a protein kinase that phosphorylates both glycogen phosphorylase (activating it) and glycogen synthase (inactivating it)
(5) The molecule cAMP is a *second messenger* in signal transduction, while the hormones binding to their receptors are the first messengers

E. **The pentose phosphate pathway**
1. The pentose phosphate pathway, also called the hexose monophosphate shunt or the phosphogluconate pathway, has two functions: to generate reducing equivalents and produce ribose 5-phosphate
2. Many biochemical reactions consume energy; biochemical systems supply this energy through molecules that carry reducing equivalents (for example, NADH and NADPH) and through molecules that carry high-energy phosphate bonds (for example, ATP, guanosine triphosphate [GTP], and UTP)
3. The pentose phosphate pathway generates reducing equivalents (conserved electrons and hydrogen atoms) in the reduced form of nicotinamide adenine dinucleotide phosphate (NADPH)
   a. NADPH is the phosphorylated form of reduced nicotinamide adenine dinucleotide (NADH)
   b. NADPH is generated in a series of reactions comprising the oxidation-reduction phase of the pentose phosphate pathway
   c. The pentose phosphate pathway occurs in the cytosol of cells and is particularly important in anabolic tissues because NADPH is needed for the biosynthesis of certain important molecules, such as fatty acids and steroids
4. Ribose 5-phosphate, another important product of the pentose phosphate pathway, is used in the synthesis of nucleotides (DNA and RNA)
   a. Ribose 5-P can be interconverted into other carbohydrates containing three, four, five, six, or seven carbon atoms
   b. A cell often uses the pentose phosphate pathway because it needs NADPH; the easy interconversion of ribose 5-P is practical because it provides other sugars that can enter glycolysis directly
5. Whether glucose proceeds through glycolysis or the pentose phosphate pathway is determined by the cell's requirement for ATP (produced in glycolysis) versus its requirement for NADPH and ribose 5-phosphate (produced in the pentose phosphate pathway)
6. The reactions of the pentose phosphate pathway occur in three main recognizable stages: the oxidation-reduction stage, the isomerization-epimerization stage, and the carbon bond cleavage-rearrangement stage (see *Pentose Phosphate Pathway*)
   a. Glucose 6-phosphate (generated in the first step of glycolysis) is oxidatively dehydrogenated at C-1, producing 6-phosphogluconolactone

Pentose Phosphate Pathway

The reactions of the pentose phosphate pathway occur in three main stages. The end products of the pathway, glyceraldehye 3-P and fructose 6-P, are intermediates in glycolysis.

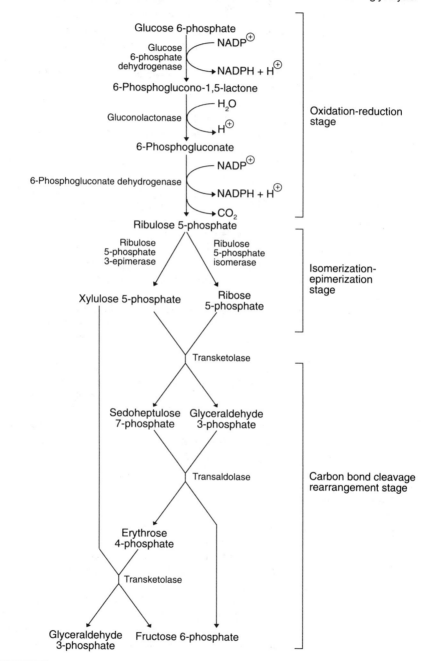

(1) *Glucose 6-P dehydrogenase* catalyzes this reaction and requires the cofactor oxidized nicotinamide adenine dinucleotide phosphate ($NADP^+$), which is reduced to $NADPH + H^+$

(2) This reaction is the rate-limiting step of the pentose phosphate pathway and is regulated by the level of $NADP^+$ in the cell

b. In the next step, which is the **commitment step** of the pathway, 6-phosphogluconolactone is hydrolyzed to 6-phosphogluconate; *gluconolactonase* catalyzes this reaction

c. The compound 6-phosphogluconate is oxidatively decarboxylated to ribulose 5-phosphate (ribulose 5-P) by *6-phosphogluconate dehydrogenase;* the cofactor $NADP^+$ is reduced to $NADPH + H^+$

d. The formation of ribulose 5-P completes the oxidation-reduction stage of the pentose phosphate pathway; all of the NADPH that will be generated by the entire pathway has been generated at this point; for each molecule of glucose entering the pathway, two molecules of NADPH are generated

e. In the isomerization-epimerization stage, ribulose 5-P is isomerized to ribose 5-phosphate (ribose 5-P) in a reaction catalyzed by *ribulose 5-P isomerase;* ribulose 5-P is also epimerized to xylulose 5-phosphate in a reaction catalyzed by ribulose 5-P epimerase

f. The remaining steps in this pathway (the carbon-carbon bond cleavage-rearrangement stage) are simply a rearrangement of carbon atoms in which three five-carbon sugars are converted to two six-carbon sugars plus a three-carbon sugar

(1) Ribose 5-P and its epimer xylulose 5-phosphate are converted, via a two-carbon transfer, to glyceraldehyde 3-phosphate (a 3-carbon sugar) and sedoheptulose 7-phosphate (a 7-carbon sugar); *transketolase* catalyzes this reaction

(2) Glyceraldehyde 3-phosphate and sedoheptulose 7-phosphate are converted, via a three-carbon transfer, to erythrose 4-phosphate (a 4-carbon sugar) and fructose 6-phosphate (a 6-carbon sugar); *transaldolase* catalyzes this reaction

(3) Erythrose 4-phosphate reacts with xylulose 5-phosphate, via a two-carbon transfer, to produce glyceraldehyde 3-phosphate and fructose 6-phosphate; *transketolase* catalyzes this reaction

F. **Gluconeogenesis**

1. **Gluconeogenesis** is the synthesis of glucose from noncarbohydrate precursors (see *Gluconeogenesis*)

    a. Lactate (an end product of anaerobic glycolysis), most amino acids (from the breakdown of dietary protein), and glycerol (an end product of triacylglycerol catabolism) are the usual noncarbohydrate precursors

    b. Lactate and amino acids are converted to either pyruvate or oxaloacetate, both of which are starting molecules for gluconeogenesis

    (1) Lactate forms pyruvate in the reverse of the reaction catalyzed by lactate dehydrogenase in glycolysis

    (2) All amino acids except leucine and lysine can be converted to either pyruvate or oxaloacetate and are **glucogenic**

Gluconeogenesis

In the illustration below, heavy arrows show the reactions unique to gluconeogenesis. Other reactions in this pathway are simply the reverse of the reactions involved in glycolysis. The noncarbohydrate precursors to gluconeogenesis are boxed; arrows indicate their entrance points.

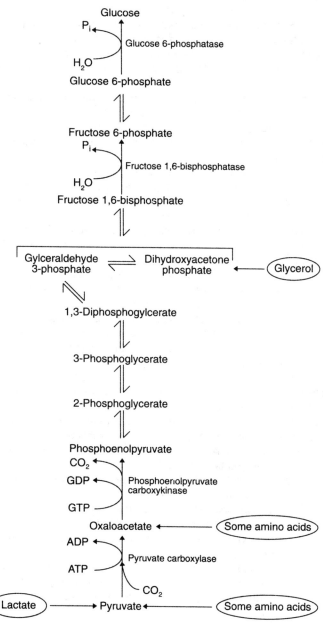

c. Glycerol is converted to dihydroxyacetone phosphate, which can enter the gluconeogenesis pathway
   (1) Glycerol is first phosphorylated to glycerol 3-phosphate (glycerol 3-P), catalyzed by glycerol kinase
   (2) Glycerol 3-P is converted to dihydroxyacetone phosphate, catalyzed by glycerol 3-P dehydrogenase
2. Some of the reactions of gluconeogenesis are simply a reversal of glycolysis and use the same enzymes as glycolysis, but other reactions use enzymes unique to gluconeogenesis; the reciprocal regulation of the separate enzyme systems enables an organism to meet its needs for both glycolysis and gluconeogenesis
3. Gluconeogenesis is necessary to maintain adequate glucose levels in the blood, especially when the diet is deficient in glucose or when the need for glucose exceeds its supply, such as during fasting
4. The liver is the major organ that performs gluconeogenesis; the kidney is a secondary site
5. The intracellular ATP concentration influences gluconeogenesis activity
   a. Low ATP levels inhibit gluconeogenesis; when ATP levels are low, acetyl coenzyme A (CoA) reacts with oxaloacetate to form citrate, and the TCA cycle proceeds to produce more ATP
   b. High ATP levels promote gluconeogenesis; a high ATP concentration decreases the activity of the TCA cycle; decreased TCA cycle activity allows oxaloacetate, the starting point for both TCA and gluconeogenesis, to be available for gluconeogenesis
6. All reactions of gluconeogenesis except the first reaction, which is catalyzed by mitochondrial pyruvate carboxylase, occur in the cytosol of cells
7. Four unique enzymes characterize gluconeogenesis: pyruvate carboxylase, phosphoenolpyruvate carboxykinase, fructose 1,6-bisphosphatase, and glucose 6-phosphatase; the remainder of the enzymes on this pathway are the enzymes of glycolysis
8. The first reaction unique to gluconeogenesis, the conversion of pyruvate to oxaloacetate, takes place in the mitochondria; the remainder of the reactions take place in the cytosol
   a. CO_2 is added to pyruvate, forming oxaloacetate; this reaction is catalyzed by pyruvate carboxylase and is the rate-limiting step of gluconeogenesis
      (1) Biotin, a vitamin, is a prosthetic group of pyruvate carboxylase and carries the activated CO_2 molecule
      (2) One molecule of ATP is hydrolyzed to adenosine diphosphate (ADP) and inorganic phosphate (P_i)
      (3) Acetyl CoA increases the activity of pyruvate carboxylase; an increase in acetyl CoA signals a need for more oxaloacetate, the molecule that brings acetyl CoA into the TCA cycle (see Chapter 3, Bioenergetics)
      (4) Mitochondrial membranes do not possess a transport system for oxaloacetate, yet oxaloacetate must be transported to the cytosol for the remaining steps of gluconeogenesis

(5) In a simple reversal of the normal TCA cycle reaction, oxaloacetate is reduced to malate (catalyzed by malate dehydrogenase); recall that malate can cross the mitochondrial membrane (see Chapter 3, Bioenergetics); once in the cytosol, malate is reoxidized to oxaloacetate and gluconeogenesis proceeds
  b. In the second reaction unique to gluconeogenesis, oxaloacetate is decarboxylated and phosphorylated to phosphoenolpyruvate; *phosphoenolpyruvate carboxykinase* catalyzes this reaction and GTP provides the high-energy phosphate group necessary for the reaction to proceed; GDP and CO_2 also are produced
  c. Phosphoenolpyruvate follows the reverse reactions of glycolysis until fructose 1,6-bisphosphate (fructose 1,6-bis-P) is formed
  d. In the third reaction unique to gluconeogenesis, fructose 1,6-bis-P is converted to fructose 6-P by the enzyme *fructose 1,6-bisphosphatase*
    (1) This reaction, which is the reverse of the phosphofructokinase (PFK) reaction in glycolysis, is highly regulated by the same molecules that regulate PFK
    (2) In glycolysis, AMP activates PFK; in gluconeogenesis, AMP inhibits fructose 1,6-bisphosphatase; AMP indicates a low cellular energy charge and a need for ATP production through glycolysis
    (3) In glycolysis, an increase in citrate inhibits PFK; in gluconeogenesis, an increase in citrate activates fructose 1,6-bisphosphatase; citrate is a product of the first TCA cycle reaction; accumulation of citrate indicates a low need for glycolysis and a high availability of oxaloacetate for gluconeogenesis
    (4) In glycolysis, fructose 2,6-bisphosphate (fructose 2,6-bisP) activates PFK; in gluconeogenesis, fructose 2,6-bisP inhibits fructose 1,6-bisphosphatase; abundant glucose triggers a hormone cascade that increases fructose 2,6-bisP, leading to an increase in glycolysis and a decrease in gluconeogenesis; low glucose levels cause low fructose 2,6-bisP levels and a decrease in PFK activity
  e. Fructose 6-P is converted to glucose 6-P by the reversal of the glycolytic phosphoglucose isomerase reaction
  f. In the fourth and last reaction unique to gluconeogenesis, glucose 6-P is converted to glucose by the hydrolysis of the phosphate group; *glucose 6-phosphatase*, which catalyzes the reaction, is present only in the liver and kidneys; thus only the liver and kidneys are capable of gluconeogenesis
9. Glycolysis and gluconeogenesis are coordinately regulated so that both processes do not operate simultaneously (futile cycling) and hence waste energy
  a. Glycolysis is controlled by regulation of its enzymes and also by the concentration of its substrate, glucose, in the blood
  b. Gluconeogenesis is controlled by regulation of its enzymes and also by the concentration of its substrates (lactate, amino acids, pyruvate, oxaloacetate, and glycerol) in the blood
10. In the liver, the conversion of lactate (produced by anaerobic glycolysis in active skeletal muscle) into glucose via gluconeogenesis is called the *Cori cycle*

a. Lactate is transported via the blood from active skeletal muscle to the liver, where it is converted to pyruvate
b. Pyruvate is converted to glucose by gluconeogenesis
c. The glucose synthesized in the liver is then transported in the blood to active skeletal muscle, where it is oxidized to produce energy for muscle contraction

Study Activities

1. Draw the structures (open chain and ring forms, where appropriate) of glucose, fructose, sucrose, ribose, glyceraldehyde, and dihydroxyacetone; circle and name the functional groups.
2. Discuss how, why, and where glucose is stored in animals.
3. Discuss how glucose storage is balanced with glucose use, and list the regulatory mechanisms of each pathway.
4. List the reactions, product(s), and cellular site(s) of the pentose phosphate pathway. Describe the biochemical reasons for the existence of this pathway.
5. Outline the reactions, product(s) and cellular site(s) of gluconeogenesis. Describe which biochemical conditions promote gluconeogenesis.
6. Define the terms listed in bold-faced type throughout this chapter.

6

Lipid Structure, Function, and Metabolism

Objectives

After studying this chapter, the reader should be able to:
- Describe the structure and function of the various categories of lipids.
- Describe the physical and chemical characteristics of membranes.
- Compare and contrast catabolism and anabolism of fatty acids.
- Outline the pathway of cholesterol biosynthesis.

I. Lipid Structure and Function

A. General information
1. Lipids are organic compounds that are insoluble, or only slightly soluble, in water and are extractable in nonpolar solvents, such as chloroform
2. Saponifiable lipids yield the salts of fatty acids when they are hydrolyzed by heat and alkali; nonsaponifiable lipids are not hydrolyzed by heat and alkali; they are the neutral fats, such as cholesterol
3. Lipids serve many physiologic functions
    a. As stored fat, they provide energy reserves
    b. As important components of both extracellular and intracellular membranes, they help separate cells and the organelles within cells
    c. As steroids, they appear in the body as cholesterol (an important component of cell membranes) and steroid hormones, such as testosterone and estradiol
    d. As *signal transducers,* they mediate the response from an extracellular signal into an intracellular biochemical event

B. Six types of lipids
1. **Fatty acids** are energy reserves; the complete oxidation of fatty acids to CO_2 and water generates adenosine triphosphate (ATP)
    a. Besides being a source of energy, fatty acids are components of other lipids, including **triacylglycerols, phospholipids,** and glycolipids; derivatives of fatty acids are hormones (such as prostaglandins) and intracellular signal transducers (such as diocylglycerol)
    b. Structurally, a fatty acid is a long hydrocarbon chain with a terminal carboxyl group, having the general structure $CH_3(CH_2)_n-COOH$ (see *Structure of a Fatty Acid,* page 94)

> **Structure of a Fatty Acid**
>
> This diagram of a typical fatty acid of unknown size shows the two different carbon labeling systems used in fatty acid nomenclature. The carbon adjacent to the carboxyl carbon is called α, and the next carbon is called β. Alternatively, carbon atoms may be numbered, beginning with the carboxyl carbon as C-1.
>
> $$H_3C-(CH_2)_n-\underset{\beta}{\overset{3}{CH_2}}-\underset{\alpha}{\overset{2}{CH_2}}-\overset{1}{C}\underset{OH}{\overset{O}{\diagdown}}$$

    c. Most higher plants and animals have fatty acids with an even number of carbon atoms (usually between 14 and 24)

    d. In labeling the carbons of a fatty acid, the carbon adjacent to the carboxyl carbon is called the α carbon and the next carbon is called the β carbon; alternatively, carbon atoms may be numbered, beginning with the carboxyl C as C-1

    e. Fatty acids are named either by naming the parent hydrocarbon and adding the suffix "oate" or by some other common name

    f. Fatty acids are either *saturated* (containing no double bonds) or *unsaturated* (containing one or more double bonds, usually in the *cis* configuration); $C_{16:0}$ denotes a 16-carbon saturated fatty acid, where 16 is the number of C atoms and 0 is the number of double bonds; a 16-carbon fatty acid with one double bond between carbon 9 and 10 is written as either $C_{16:1}\Delta^9$ or $C_{16:1}(9)$, where Δ^9 or (9) denotes the position of the unsaturated bond

2. **Triacylglycerols,** also called triglycerides, are the storage form of fatty acids; they are located mainly in the cytoplasm of adipose cells and liver cells

    a. Structurally, a triacylglycerol is three (*tri*) fatty acids linked by their *acyl* groups in an ester linkage to *glycerol* (see *Structure of a Triacylglycerol*)

    b. The carbon atoms of glycerol are numbered as *s*tereospecific *n*umbers (sn) 1, 2, and 3 starting from the top when glycerol is drawn in a Fischer projection with the C-2 hydroxyl group projecting to the left

    c. If the fatty acid at C-1 differs from that at C-3, then C-2 is asymmetric (see Chapter 1, Basics of Biochemistry); naturally occurring triacylglycerols have the L configuration

3. *Glycerophospholipids* are *sn* glycerol 3-phosphate with fatty acids esterified at the C-1 and C-2 positions of glycerol and with various other groups derived from polar alcohols esterified at the phosphate group on C-3

    a. When only one fatty acid is esterified to glycerol 3-phosphate *sn* 1 or *sn* 2, *lysophosphatidate* is formed; when a second fatty acid is joined to lysophosphatidate, *phosphatidate* is formed; glycerophospholipid is formed when a polar alcohol is esterified to the phosphate group of phosphatidate

    b. Glycerophospholipids vary depending upon the fatty acids esterified to the glycerol backbone and the group esterified to the phosphate

    c. The fatty acids at C-1 are usually saturated, while those at C-2 are usually unsaturated

Structure of a Triacylglycerol

The backbone of the triacylglycerols is the glycerol molecule (shown boxed). Triacylglycerols are formed from the esterification of fatty acids with glycerol; if the fatty acid at C-1 is different from the fatty acid at C-3, then the molecule is asymmetric at C-2.

Glycerol

$$\begin{array}{c} ^1CH_2OH \\ | \\ HO-^2CH \\ | \\ ^3CH_2OH \end{array}$$

Triacylglycerol

$$\begin{array}{c} \quad\quad\quad\quad O \\ \quad\quad\quad\quad \| \\ \quad\quad\quad ^1CH_2-O-C-R_1 \\ O\quad\quad | \\ \|\quad\quad ^2CH \quad\quad O \\ R_2-C-O-\quad\quad\quad \| \\ \quad\quad\quad ^3CH_2-O-C-R_3 \end{array}$$

    d. *Phosphatidylinositol*, so named because the carbohydrate inositol is bound to phosphatidate, generates important signal transducers; the phospholipids phosphatidylcholine, phosphatidylethanolamine, and phosphatidylserine are all named for the X group attached to the phosphate of phosphatidate

    e. Glycerophospholipids are *amphipathic* because of their polar "head group" (charged phosphate or substituents bound to phosphate) and their two nonpolar fatty acid "tail groups"

4. *Sphingolipids* contain sphingosine as the backbone to which other groups are attached
    a. Sphingosine is a C_{18} amino alcohol with a single unsaturated bond in the C_{18} chain
    b. Fatty acids form an amide linkage with the amino group of sphingosine, forming a *ceramide*
    c. Phosphate can be linked to the primary alcohol of a ceramide to form a sphingophospholipid; when phosphocholine is attached, the compound formed is sphingomyelin, which is the most abundant sphingolipid and the only sphingolipid found in membranes (see *Structure of Sphingomyelin,* page 96)

5. *Glycolipids* contain a carbohydrate molecule bound at *sn* 3 to a lipid alcohol group through a glycosidic bond
    a. Glycolipids are classified as either glycerides (if glycerol is the backbone) or glycosphingolipids (if sphingosine is the backbone); most naturally ocurring glycolipids are glycosphingolipids
    b. Two important categories of glycosphingolipids are cerebrosides and gangliosides; *cerebrosides* are ceramides with either glucose or galactose attached to the primary sphingosine hydroxyl group; *gangliosides* are complex cerebrosides that contain many carbohydrates (oligosaccharides), linked by a glucose residue, attached to sphingosine.

6. *Steroids* include various compounds with the same general structure: a fused-ring system consisting of three six-membered rings plus one five-membered ring
    a. The most important steroids include sex hormones and cholesterol
    b. *Cholesterol,* an important component of membranes in animals, is not present in plants

> **Structure of Sphingomyelin**
>
> Sphingomyelin is the most abundant sphingolipid; it is formed from a sphingosine backbone to which a phosphocholine and a palmitic acid residue have been added.
>
> [Structural diagram showing phosphocholine head group: $CH_3-N^+(CH_3)(CH_3)-CH_2-CH_2-O-P(=O)(O^-)-O-$, attached to sphingosine backbone $CH_2-C(NH-)-C(OH)H-CH=CH-(CH_2)_{12}-CH_3$, with palmitate residue $O=C-(CH_2)_{14}-CH_3$ attached via the NH.]

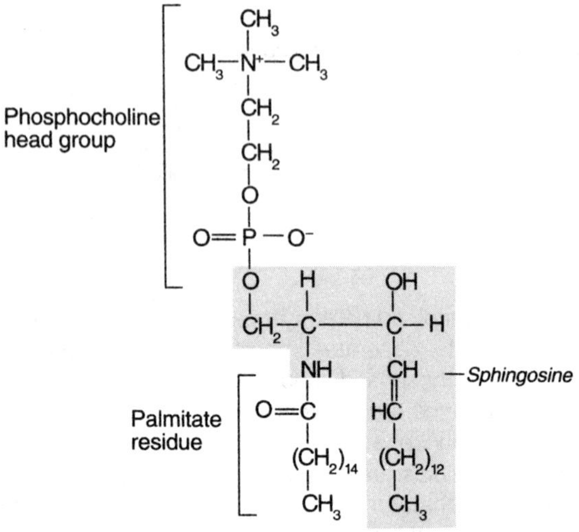

    c. Structurally, cholesterol contains 27 carbon atoms in four fused rings, with an eight-member branched hydrocarbon chain at C-17, a hydroxyl group at C-3, two nonring methyl groups designated C-18 and C-19, and an unsaturation between C-5 and C-6

    d. Cholesterol is a precursor of the steroid hormones pregnenolone (C_{21}) and progestogens (C_{21}); progestogens in turn are precursors of glucocorticoids (C_{21}), mineralocorticoids (C_{21}), and androgens (C_{19}); androgens are precursors of estrogens (C_{18})

    e. Bile acids, which are polar derivatives of cholesterol, are products of cholesterol catabolism; two bile acids are glycocholate and taurocholate

    f. Bile acids are synthesized in the liver, stored and concentrated in the gallbladder, and released into the small intestine to solubilize dietary lipids

II. Membranes

A. General information

1. All cells have a cell membrane (also known as the plasma membrane); in addition, eukaryotic cells have membranes that enclose their various organelles
2. Membranes are dynamic structures that control the space they enclose by selectively excluding or including certain molecules
      a. Membranes may contain biologically active molecules that are **receptors** for messenger molecules, such as hormones and proteins

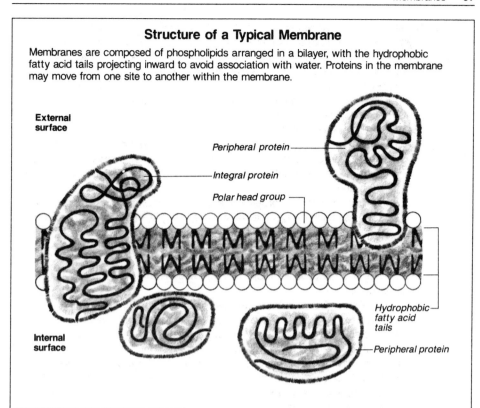

Structure of a Typical Membrane

Membranes are composed of phospholipids arranged in a bilayer, with the hydrophobic fatty acid tails projecting inward to avoid association with water. Proteins in the membrane may move from one site to another within the membrane.

      b. Membranes may contain molecules that act as signal transducers and other molecules that serve as markers differentiating one cell type from another
   3. Lipids provide the structural components of membranes and account for the hydrophobicity of membranes; phospholipids (glycerophospholipids and sphingophospholipids) and cholesterol are the major lipid components of membranes
   4. Proteins constitute from 20% to 80% of membranes, depending upon the particular membrane; intracellular (organelle) membranes have a higher protein content than cellular membranes
   5. Carbohydrates, as glycoproteins or glycolipids, are a minor component of membranes

B. Membrane composition
   1. Membrane phospholipids are arranged in a bilayer structure (see *Structure of a Typical Membrane*)
      a. The polar head groups constitute the two outer surfaces exposed to the aqueous environment, one facing the exterior of the cell or organelle and the other facing the interior
      b. The fatty acid tail groups project inward and form a nonpolar environment shielded from water

2. Membranes are fluid rather than static, a characteristic that permits a cell or organelle to both adjust its shape and move; the degree of fluidity in a membrane depends on the type of fatty acids in the phospholipid bilayer
   a. Long-chain saturated fatty acids form an extended, more rigid structure
   b. Short-chain fatty acids and unsaturated fatty acids enhance membrane fluidity because the hydrocarbon tail of an unsaturated fatty acid bends at the double bond, allowing a less ordered (more fluid) structure
3. Membranes are asymmetric
   a. The carbohydrate residues of glycolipids or glycoproteins are oriented so that they project to the exterior surface of the plasma membrane
   b. Proteins in the membrane may move from location to location within the membrane
   c. Peripheral, or extrinsic, proteins are on the membrane surface and are easily dissociated from the membrane; integral, or intrinsic, proteins are difficult to dissociate from the membrane; they are embedded within the membrane and may project partially into the lipid bilayer or completely transverse the membrane
4. Membranes use noncovalent molecular interactions to achieve and maintain their structure
   a. The nonpolar fatty acid tails are hydrophobic (repelled by water) and attracted to each other by van der Waals attractive forces
   b. The polar head groups are attracted to water through hydrogen bonding and ionic interactions with water molecules
5. The fluid mosaic model of membranes accounts for the mobility of membrane proteins and the various degrees of fluidity observed in membranes
   a. Membrane lipids serve two main functions: they provide a permeability barrier, especially against hydrophilic molecules, and they accommodate integral proteins (the conformational bend produced by an unsaturated fatty acid may permit protein insertion)
   b. Both lipids and proteins are free to move and diffuse within the membrane
6. Molecules can move through a membrane by a variety of mechanisms; the type of transport depends on the specific molecule and the specific membrane
   a. In *passive transport,* a molecule moves across a membrane from a region of higher concentration to a region of lower concentration; because the molecule is not moving against a concentration gradient, the cell expends no energy
      (1) *Simple diffusion* is a type of passive transport in which the molecule moves through an opening or pore in the membrane following the concentration gradient; small molecules, such as water or oxygen, move directly through cell membranes via simple diffusion
      (2) In *facilitated diffusion,* the molecule to be transported binds to a carrier protein, which then changes conformation to move the molecule across the membrane
   b. In *active transport,* a molecule moves across the membrane from a region of low concentration to a region of high concentration; moving a molecule against its concentration gradient requires energy expenditure

(1) The active transport of Na^+ to the outside of a cell and K^+ to the inside of the cell are examples of ion movement against the normal concentration gradient
(2) To facilitate the movement of these ions, a single type of membrane protein both hydrolyzes ATP for energy and transports the ion against its concentration gradient; the ion movement performed by this membrane protein is called the *sodium-potassium ion pump*

III. Lipid Metabolism

A. General information
1. Lipids or their component molecules are transported throughout the body in various forms
   a. Lipids can be found in blood plasma complexed to various proteins (plasma lipoproteins), such as chylomicrons, very-low-density lipoproteins (VLDL), low-density lipoproteins (LDL), and high-density lipoproteins (HDL)
   b. Fatty acids can be bound to the plasma protein albumin
2. The complete oxidation of a fatty acid yields approximately twice as much ATP than complete oxidation of a carbohydrate; in animals, stored fat contains a much greater reservoir of energy than a comparable weight of stored glycogen
3. Catabolism of fatty acids from triacylglycerols provide the bulk of the energy animals extract from lipids
   a. Fatty acids are catabolized through a process called β oxidation, so named because oxidation occurs at the β carbon of the fatty acid
   b. Most of the reactions of β oxidation occur in the mitochondrial matrix; only the initial reaction (the activation of fatty acid) occurs in the cytosol

B. Catabolism of fatty acids
1. Triacylglycerols, stored in the cytoplasm of adipose and liver cells, are first hydrolyzed into fatty acids and glycerol through the action of specific *lipases* that hydrolyze bonds at *sn* 1, *sn* 2, and *sn* 3 positions of triacylglycerols; the glycerol released is converted ultimately to glyceraldehyde 3-phosphate, which may be used in either glycolysis or gluconeogenesis
2. Fatty acids released by lipases are first "activated" by forming a thioester bond with coenzyme A (CoA) forming acyl CoA; *acyl CoA synthetase* catalyzes this reaction using two high-energy bonds from the hydrolysis of ATP to adenosine monophosphate (AMP) and inorganic pyrophosphate (PP_i)
   a. The activated fatty acids are transported from the cytosol across the outer mitochondrial membrane and into the intermembrane space
   b. *Carnitine,* a derivative of the amino acid lysine, transports the fatty acids across the inner mitochondrial membrane and into the matrix
      (1) At the outer surface of the inner mitochondrial membrane, the fatty acid releases CoA, forms a bond with carnitine, and crosses the inner surface of the inner mitochondrial membrane as acyl carnitine

β Oxidation of a Fatty Acid

Before β oxidation begins, a fatty acid is first activated by forming a thioester bond to coenzyme A (CoA); the activated fatty acid is then transported from the cytosol and into a mitochondrion, where the process of β oxidation takes place.

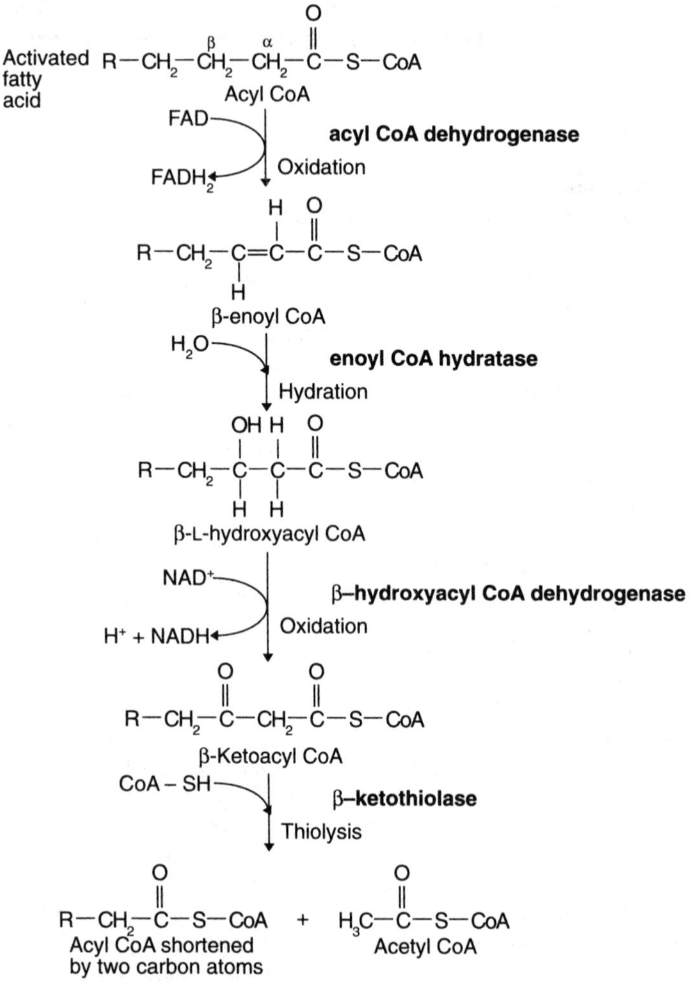

Activated fatty acid:

$$R-CH_2-\overset{\beta}{CH_2}-\overset{\alpha}{CH_2}-\overset{O}{\underset{\|}{C}}-S-CoA$$

Acyl CoA

↓ acyl CoA dehydrogenase, Oxidation (FAD → FADH$_2$)

$$R-CH_2-\underset{H}{\overset{H}{C}}=\overset{}{C}-\overset{O}{\underset{\|}{C}}-S-CoA$$

β-enoyl CoA

↓ enoyl CoA hydratase, Hydration (H$_2$O)

$$R-CH_2-\underset{H}{\overset{OH}{C}}-\underset{H}{\overset{H}{C}}-\overset{O}{\underset{\|}{C}}-S-CoA$$

β-L-hydroxyacyl CoA

↓ β-hydroxyacyl CoA dehydrogenase, Oxidation (NAD$^+$ → H$^+$ + NADH)

$$R-CH_2-\overset{O}{\underset{\|}{C}}-CH_2-\overset{O}{\underset{\|}{C}}-S-CoA$$

β-Ketoacyl CoA

↓ β-ketothiolase, Thiolysis (CoA–SH)

$$R-CH_2-\overset{O}{\underset{\|}{C}}-S-CoA \quad + \quad H_3C-\overset{O}{\underset{\|}{C}}-S-CoA$$

Acyl CoA shortened by two carbon atoms   Acetyl CoA

(2) On the matrix side of the inner mitochondrial membrane, the fatty acid once again complexes with another CoA to re-form acyl CoA and releases carnitine

(3) The enzymes carnitine acyltransferase I (on the outer surface of the inner mitochondrial membrane) and carnitine acyltransferase II (on the matrix side of the inner mitochondrial membrane) catalyze these reactions

3. Inside the mitochondrial matrix, the fatty acid is catabolized in an energy-producing process called β oxidation (see β *Oxidation of a Fatty Acid*)
   a. The single bond between the α and the β carbon of acyl CoA is oxidized to a *trans* double bond, forming β-enoyl CoA; *acyl CoA dehydrogenase,* the enzyme that catalyzes this reaction, requires oxidized flavin adenine dinucleotide (FAD) as a coenzyme; FAD is reduced to $FADH_2$ and then enters the respiratory chain to produce 2 ATP molecules via oxidative phosphorylation (see Chapter 3, Bioenergetics)
   b. A molecule of water is added to the double bond, forming β-L-hydroxyacyl CoA; the enzyme *enoyl CoA hydratase* catalyzes the reaction
   c. The new hydroxyl group of β-L-hydroxyacyl CoA is oxidized to a carbonyl group, forming β-ketoacyl CoA; this reaction catalyzed by *β-hydroxyacyl CoA dehydrogenase* requires nicotinamide adenine dinucleotide (NAD^+) as a cofactor; NAD^+ is reduced to NADH and then enters the respiratory chain, producing 3 ATP molecules via oxidative phosphorylation
   d. The enzyme *β-ketothiolase,* also called acetyl CoA acetyltransferase or thiolase, catalyzes the cleavage of β-ketoacyl CoA; this reaction requires a molecule of acetyl CoA and produces acetyl CoA and an acyl CoA that is two carbons shorter than the original fatty acid molecule
   e. The shortened fatty acid chain repeats the β-oxidation pathway until the fatty acid is completely oxidized to acetyl CoA
4. Each cycle of β oxidation (except the last) produces one molecule each of $FADH_2$, NADH + H^+, and acetyl CoA; the last cycle produces two molecules of CoA because a four-carbon fatty acid is cleaved in half
5. As an example, the catabolism of palmitate ($C_{16:0}$) generates 129 molecules of ATP as it is degraded in seven rounds of β oxidation
   a. Eight molecules of acetyl CoA are produced directly through β oxidation
   b. In the tricarboxylic acid (TCA) cycle and respiratory chain, each acetyl CoA generates 12 molecules of ATP (see Chapter 3, Bioenergetics); therefore $8 \times 12 = 96$ molecules of ATP are produced
   c. Seven molecules of $FADH_2$ are produced during β oxidation of the entire palmitate molecule; in the respiratory chain each $FADH_2$ generates two molecules of ATP; therefore $7 \times 2 = 14$ molecules of ATP are produced from $FADH_2$
   d. Seven molecules of NADH are produced during β oxidation of the entire palmitate molecule; in the respiratory chain each NADH generates 3 molecules of ATP; therefore $7 \times 3 = 21$ molecules of ATP are produced
   e. Two high-energy bonds (ATP converted to AMP) are expended in activation of the fatty acid; therefore $96 + 14 + 21 - 2 = 129$ molecules of ATP are produced by the complete oxidation of one palmitate molecule
6. Catabolism of unsaturated fatty acids requires additional enzymes
   a. Activation of unsaturated fatty acids and their transport into mitochondria occur by the same processes as for saturated fatty acids
   b. β oxidation proceeds as for saturated fatty acids until the site of unsaturation is reached; a different reaction then occurs because the enzyme acyl CoA dehydrogenase does not recognize a double bond as its substrate
   c. An *isomerase* then shifts the unsaturation to the adjacent carbon and also converts it from a *cis* to *trans* configuration

d. The *trans* configuration of the double bond is a normal substrate for the enzyme enoyl CoA hydratase and β oxidation proceeds normally
   e. If a fatty acid containing more than one unsaturated (polyunsaturated) bond is oxidized, *epimerase* is also needed
7. In certain situations, acetyl CoA is used preferentially over glucose as an energy source; during fasting or starvation, fat is mobilized from adipose tissue and metabolized for energy; in diabetes, glucose is not available for glycolysis because a shortage of insulin prevents glucose entry into cells
   a. Both fasting and diabetes cause acetyl CoA (from the β oxidation of fatty acids) to be used preferentially over glucose as an energy source; when acetyl CoA is the primary fuel oxidized, *ketone bodies* are produced
   b. Acetyl CoA is present in higher amounts than oxaloacetate, the compound it initially joins in the TCA cycle (see Chapter 3, Bioenergetics); excess acetyl CoA forms *acetoacetate*, which is spontaneously decarboxylated to *acetone*, and β-*hydroxybutyrate*; these three compounds are called ketone bodies
   c. Acetoacetate and β-hydroxybutyrate, previously thought to be nonfunctional byproducts, are used as energy sources by the heart and, in starvation or diabetes, by the brain
8. In normal, healthy states, any acetyl CoA not needed for immediate energy production is used to synthesize fatty acids, which are storage forms of energy for later use

C. Fatty acid synthesis
1. Fatty acid synthesis takes place in the cytosol of liver cells; two preliminary steps are necessary before fatty acid synthesis can begin
   a. Fatty acid synthesis begins with acetyl CoA in the cytosol; acetyl CoA must somehow leave the mitochondria and be transferred to the cytosol
   b. Acetyl CoA is produced in the mitochondria, both from β oxidation of fatty acids and from pyruvate through the action of the pyruvate dehydrogenase complex (see Chapter 3, Bioenergetics)
   c. Acetyl CoA does not readily cross the mitochondrial membrane; instead, it reacts with oxaloacetate to form citrate, catalyzed by citrate synthase (a normal step in the TCA cycle); this is the first preliminary step to fatty acid synthesis
   d. Citrate is transported from the mitochondria into the cytosol where it crosses the outer mitochondrial membrane and reacts with CoA and ATP, forming acetyl CoA, oxaloacetate, adenosine diphosphate (ADP), and inorganic phosphate (P_i); this reaction is catalyzed by citrate lyase
   e. CO_2, in the form of bicarbonate ion (HCO_3^-) is added to acetyl CoA to form malonyl CoA ; this reaction, catalyzed by *acetyl CoA carboxylase*, is the second preliminary step to fatty acid synthesis
      (1) This reaction, the commitment step and the rate limiting step of fatty acid synthesis, requires ATP, which is hydrolyzed to ADP and P_i; manganese ions (Mn^{2+}) and biotin are cofactors
      (2) Acetyl CoA carboxylase is allosterically activated by increased concentrations of citrate and isocitrate, TCA cycle components that signal that energy is available to be stored as fatty acids

(3) Acetyl CoA carboxylase is allosterically inhibited by the ultimate product of fatty acid synthesis, palmitoyl CoA
   f. The synthesis of fatty acids is the successive addition of two-carbon units to the malonyl CoA starting molecule; the usual end product of fatty acid synthesis is palmitate, C_{16}
2. The reactions that convert malonyl CoA to a fatty acid are catalyzed by *fatty acid synthase*, which in bacteria is a complex of different proteins; in higher organisms, fatty acid synthase is a large multienzyme complex contained in one protein (see *Biosynthesis of a Fatty Acid*, page 104)
   a. Intermediate compounds in fatty acid synthesis are bound to an *acyl carrier protein (ACP)*, similar to the linkage of intermediate compounds to CoA in fatty acid catabolism
   b. To begin the reaction, acetyl CoA binds to a part of ACP; the bonding of acetyl CoA to ACP to form acetyl ACP is catalyzed by *ACP acetyltransferase* (not shown in diagram)
   c. Malonyl, as malonyl CoA, is transferred to a different site on the ACP forming malonyl-ACP and CoA; the reaction is catalyzed by *ACP-malonyltransferase* (not shown in diagram)
   d. The enzyme β-*ketoacyl-ACP-synthase* catalyzes the condensation between one molecule each of malonyl-ACP and acetyl-ACP, to form one molecule of acetoacetyl-ACP and release one molecule each of ACP and CO_2; the carbons from the acetyl-ACP are always the two terminal carbons of the growing fatty acid chain
   e. In the first of two reduction reactions, acetoacetyl-ACP is reduced to β-hydroxybutyryl-ACP; the reaction, catalyzed by β-*ketoacyl-reductase*, requires the cofactor reduced nicotinamide adenine dinucleotide phosphate (NADPH) and H^+, which is subsequently oxidized to $NADP^+$
   f. β-hydroxybutyryl-ACP is dehydrated to crotonyl-ACP in a reaction catalyzed by β-*hydroxyacyl-ACP dehydratase*
   g. In the second reduction reaction, crotonyl-ACP is reduced to butyryl-ACP; this reaction, catalyzed by β-*enoyl-ACP-reductase*, requires the cofactor NADPH, which is subsequently oxidized to $NADP^+$
3. In subsequent rounds of fatty acid synthesis, two-carbon units that are derived from malonyl-ACP (a three-carbon compound) are added to the growing fatty acid chain until after seven cycles, the sixteen-carbon compound palmitoyl-ACP is formed; in the final step, palmate is hydrolyzed from ACP by palmitoyl-ACP deacylase and may be transferred to CoA
4. Fatty acids containing an odd number of carbon atoms are synthesized by the same reactions; however, propionyl CoA (a three-carbon compound) is used in place of acetyl CoA for condensation with malonyl-ACP
   a. The enzyme ACP-acyltransferase can transfer any acyl groups, not just acetyl groups
   b. The first condensation reaction yields a five-carbon intermediate and CO_2
5. Production of fatty acids longer than palmitate involves the addition of acetyl units; malonyl CoA is the donor of the two-carbon units
6. Production of unsaturated fatty acids occurs in mitochondria and requires NADH, O_2, and cytochrome b_5
7. Humans cannot synthesize *essential* fatty acids and must therefore obtain them from the diet

Biosynthesis of a Fatty Acid

The reaction sequence shown here is the early stages of biosynthesis of palmitic acid ($C_{16:0}$), the usual end product of fatty acid synthesis. In subsequent reaction sequences, two-carbon units derived from malonyl-ACP (acyl carrier protein) are added to the growing fatty acid chain until palmitoyl-ACP is formed.

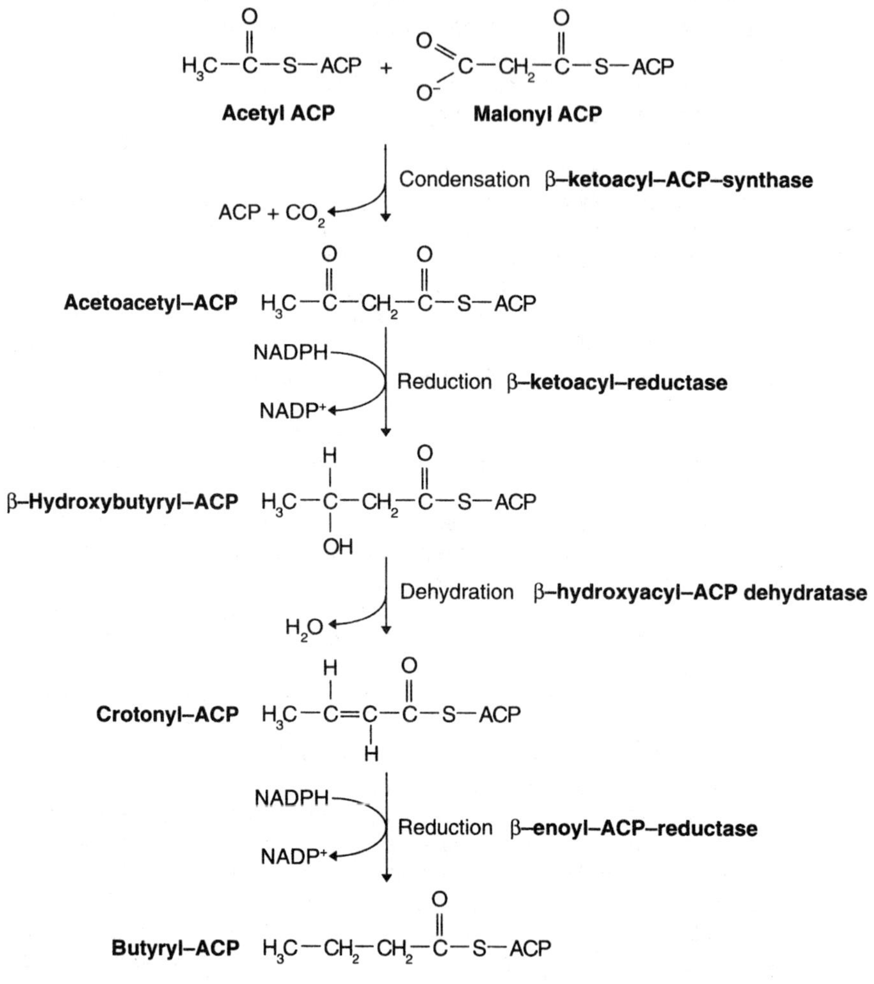

$$H_3C-\underset{\underset{O}{\|}}{C}-S-ACP \quad + \quad ^{-}O-\underset{\underset{O}{\|}}{C}-CH_2-\underset{\underset{O}{\|}}{C}-S-ACP$$

Acetyl ACP   **Malonyl ACP**

↓ Condensation β-ketoacyl-ACP-synthase
ACP + CO_2 ←

Acetoacetyl–ACP $H_3C-\underset{\underset{O}{\|}}{C}-CH_2-\underset{\underset{O}{\|}}{C}-S-ACP$

NADPH ↘
↓ Reduction β-ketoacyl-reductase
$NADP^+$ ↙

β-Hydroxybutyryl–ACP $H_3C-\underset{OH}{\overset{H}{C}}-CH_2-\underset{\underset{O}{\|}}{C}-S-ACP$

↓ Dehydration β-hydroxyacyl-ACP dehydratase
H_2O ←

Crotonyl–ACP $H_3C-\underset{H}{\overset{H}{C}}=C-\underset{\underset{O}{\|}}{C}-S-ACP$

NADPH ↘
↓ Reduction β-enoyl-ACP-reductase
$NADP^+$ ↙

Butyryl–ACP $H_3C-CH_2-CH_2-\underset{\underset{O}{\|}}{C}-S-ACP$

    a. Linoleate ($C_{18:2}$), linolenate ($C_{18:3}$), and arachidonate ($C_{20:4}$) are essential fatty acids
    b. Linoleate is considered the only true essential fatty acid, because linolenate and arachidonate can be produced from it

D. **Cholesterol synthesis**
    1. Cholesterol synthesis occurs in the cytosol of cells; all 27 carbon atoms in cholesterol are derived from acetate, in the form of acetyl CoA

2. The two-carbon compound acetyl CoA is converted into the five-carbon (C_5) compound isopentenyl pyrophosphate
   a. Acetyl CoA is first complexed with acetoacetyl CoA, forming the C_6 compound 3-hydroxy-3-methylglutaryl (HMG) CoA
   b. HMG CoA is converted to mevalonate by *HMG CoA reductase;* this is the commitment step of cholesterol synthesis
   c. In three reactions that each use ATP, mevalonate is converted into isopentenyl pyrophosphate, a C_5 compound
   d. Isopentenyl pyrophosphate isomerizes to dimethylallyl pyrophosphate; two molecules of dimethylallyl pyrophosphate condense to form the C_{10} compound geranyl pyrophosphate
   e. Geranyl pyrophosphate condenses with one molecule of dimethylallyl pyrophosphate to form the C_{15} compound farnesyl pyrophosphate
   f. Two molecules of farnesyl pyrophosphate condense to form the C_{30} compound squalene
   g. Squalene is oxidized, forming an epoxide, which cyclizes to form lanosterol
   h. Three carbon atoms are removed from lanosterol, forming the C_{27} compound cholesterol

Study Activities

1. Draw the structure and state the function(s) of each of the major categories of lipids.
2. Sketch a typical membrane and label its components.
3. Describe the reactions of β oxidation of fatty acids.
4. Account for the ATP used and produced in the oxidation of palmitate.
5. Describe the structure, function, and formation of ketone bodies.
6. Describe the reactions of fatty acid synthesis.
7. Describe the reactions of cholesterol synthesis.
8. Define the terms listed in bold-faced type throughout this chapter.

7

Transmission of Genetic Information

Objectives

After studying this chapter, the reader should be able to:
- Define replication, transcription, and translation, and describe the interrelationship among them.
- Describe, compare, and contrast the composition of DNA, RNA, and protein.
- List the cellular location for replication, transcription, translation, and post-translational modifications.
- Describe the processes of replication, transcription, and translation in prokaryotes, and distinguish the differences in these processes in prokaryotes and eukaryotes.
- State the purpose of DNA repair mechanisms and describe excision repair.
- State the purpose of post-translational modification of protein and describe several modification types.

I. Components of Genetic Information

A. General information
1. The three major components in the transmission of genetic information are **deoxyribonucleic acid (DNA),** ribonucleic acid (RNA), and protein
2. Most DNA is located in the nuclear region of prokaryotic cells and in the nucleus of eukaryotic cells; DNA has two primary functions in cells
   a. DNA determines which proteins a cell will produce; the sequence of nitrogenous bases that compose DNA serve as a **template** (blueprint) for all the proteins the cell will synthesize
   b. When a cell divides, DNA ensures the conservation of genetic information; the DNA duplicates itself in a multistep process called **replication**
3. Cells have three major types of RNA, each of which has a different function in protein synthesis
   a. Messenger RNA (mRNA) is produced directly from a DNA template; the formation of mRNA from a DNA template is called **transcription;** in eukaryotic cells, newly formed mRNA leaves the nucleus to direct protein synthesis at the ribosomes; in prokaryotes, mRNA begins to direct protein synthesis before transcription is even complete
      (1) In contrast to the vast majority of cells that transfer genetic information in the sequence DNA → RNA, retroviruses can transfer genetic information in the sequence RNA → DNA, a process that is the reverse of normal transcription

(2) A well-known retrovirus is the human immunodeficiency virus (HIV), the putative cause of acquired immunodeficiency syndrome (AIDS)
  b. Ribosomal RNA (rRNA) is an integral part of ribosomes, particles of RNA and protein that are the sites of protein synthesis in all organisms
  c. Transfer RNA (tRNA), which is found in cytosol, transfers an amino acid to the ribosome for protein synthesis; **translation** is the process by which ribosomes "read" an attached mRNA molecule, then accept the tRNA bearing the correct amino acids needed for protein synthesis
4. Proteins are polymers of amino acids; the distinct sequence of amino acids in a protein confers its unique properties and function (See Chapter 4, Protein Function and Metabolism)
  a. Cells express genetic information through protein synthesis; most cells express genetic information through the sequence

$$\text{DNA} \xrightarrow[\text{transcription}]{\text{replication}} \text{RNA} \xrightarrow{\text{translation}} \text{protein}$$

  b. Proteins are the manifestation of the **genetic code** of DNA
5. Fidelity in the transmission of genetic information is important to ensure that progeny receive the correct complement of genetic instructions from each parent; the genetic information must not have any missing or added data and must be free of errors.

B. DNA composition
1. DNA is composed of repeating units of a nitrogenous base linked to the five-carbon sugar β-D-2-deoxyribose, which, in turn, is linked to the sugar before and after it by **phosphodiester** bonds; each deoxyribose carries a nitrogenous base

Phosphate ester   β-D-2-Deoxyribose

  a. The sugar carbons are numbered with a prime (') sign to differentiate them from the carbons in nitrogenous bases
  b. Because all carbohydrates are chiral molecules, the D designation in β-D-2-deoxyribose indicates the absolute configuration of the highest numbered chiral center when written in a Fischer projection formula (see Chapter 1, Basics of Biochemistry)
  c. The β designation refers to the above-ring position of the hydroxyl (OH) group attached to the anomeric carbon (C-1') (see Chapter 1, Basics of Biochemistry)
  d. The designation 2-deoxyribose indicates that DNA lacks an oxygen atom at the 2' position

e. The nitrogenous bases attached to deoxyribose are heterocyclic molecules classified as either **purines** or **pyrimidines** depending on their structure
   (1) The two purines are adenine (A) and guanine (G), both of which are found in DNA and RNA

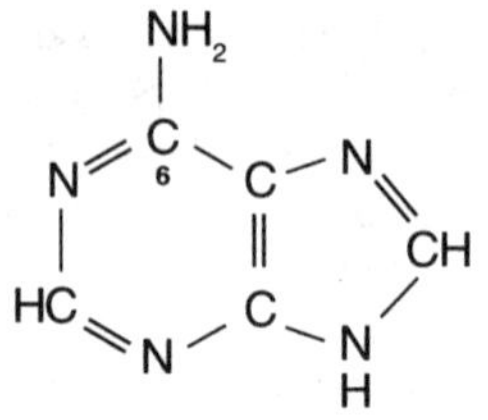

Adenine

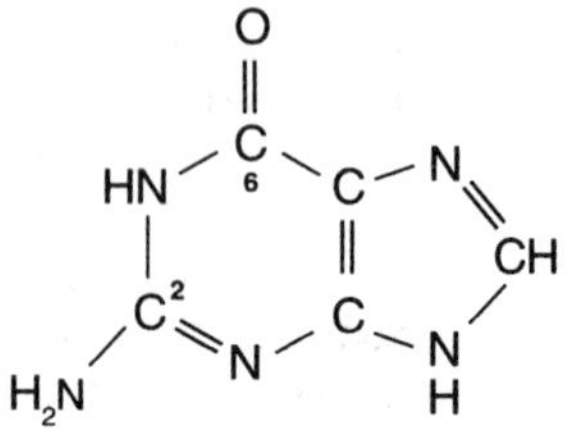

Guanine

   (2) The three pyrimidines are cytosine (C), which is found in both DNA and RNA; thymine (T), which is found in DNA only; and uracil (U), which is found in RNA only

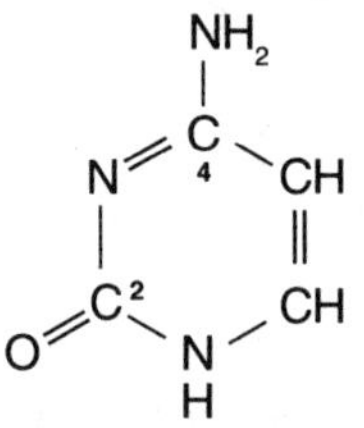

Cytosine (C)

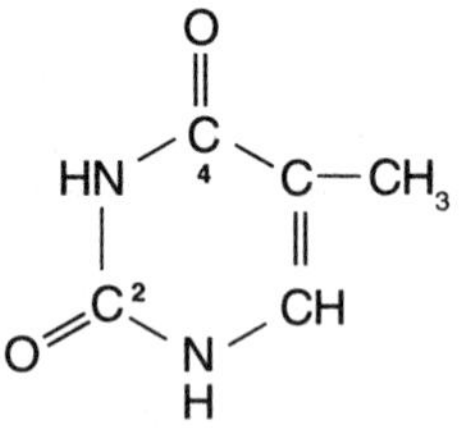

Thymine (T)

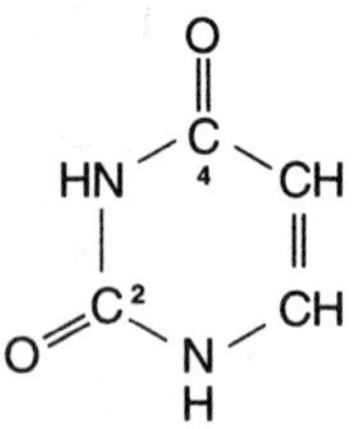

Uracil (U)

2. A **nucleoside** is a molecule composed of a purine or pyrimidine bonded to a pentose sugar through a β-N-glycosidic bond; a glycoside is a compound in which one or more sugars are bound to another molecule
   a. Purines are linked through their N-9 to the C-1' of the carbohydrate

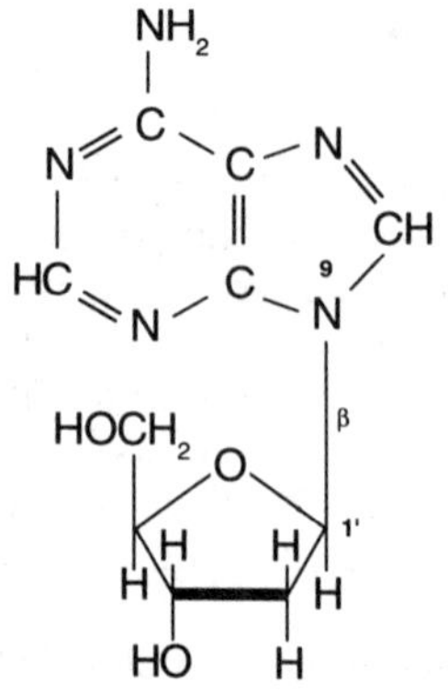

The nucleoside deoxyadenosine

Nucleotide of Adenosine

Attachment of a mono-, di-, or triphosphate group to the hydroxyl group of the 5' carbon of a nucleoside converts the molecule to a nucleotide. The formula below shows adenosine triphosphate (ATP), the nucleotide of adenosine with three phosphate groups attached.

$$^-O-\underset{\underset{O^-}{|}}{\overset{\overset{O}{\|}}{P}}-O-\underset{\underset{O^-}{|}}{\overset{\overset{O}{\|}}{P}}-O-\underset{\underset{O^-}{|}}{\overset{\overset{O}{\|}}{P}}-O-CH_2-\text{(ribose)}-\text{adenine}$$

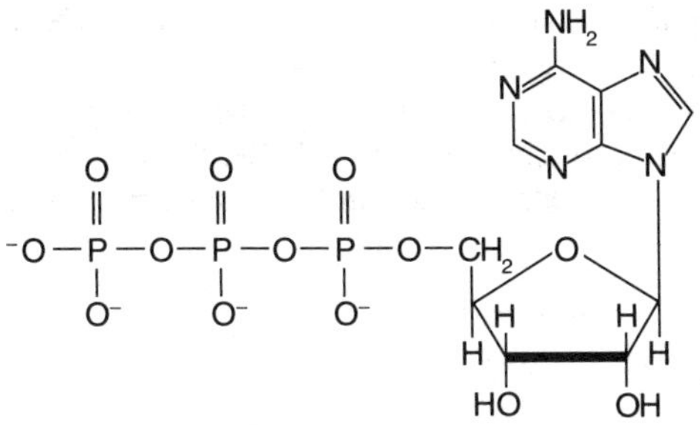

b. Pyrimidines are linked through their N-1 to the C-1' of the carbohydrate

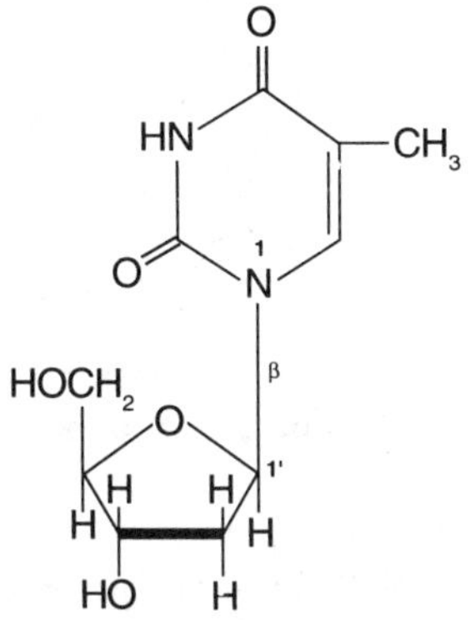

The nucleoside deoxythymidine

c. Nucleosides are named by adding the suffix "-osine" or "-idine" to the name of the base (for example, deoxyadenosine, deoxyguanosine, deoxycytidine, and deoxythymidine are the nucleosides of the sugar deoxyribose; adenosine, guanosine, cytidine, and uridine are nucleosides of the sugar ribose)

3. A **nucleotide** is a nucleoside with phosphate attached to the 5' carbon; one, two, or three phosphate groups may be attached as sequential phosphate esters of the 5' carbon (see *Nucleotide of Adenosine*)

a. Nucleotides are named as the nucleoside with the number of phosphate groups attached to the carbohydrate indicated; for example, the nucleotide of deoxyadenosine is named as deoxyadenosine monophosphate (dAMP), deoxyadenosine diphosphate (dADP), or deoxyadenosine triphosphate (dATP), depending on whether one, two, or three phosphate groups, respectively, are attached

b. Nucleotides are commonly indicated by acronyms in which an initial lowercase "d" indicates the deoxynucleotide; for example, AMP is adenosine monophosphate, and dAMP is deoxyadenosine monophosphate

4. Individual deoxynucleotides are linked to form a DNA strand; DNA occurs in nature as a pair of single DNA strands connected by hydrogen bonding between specific nitrogenous bases of each strand
   a. In a single strand of DNA, each nucleotide is linked to the next nucleotide by a phosphodiester bond at its 3'-OH group and to the preceding nucleotide by a phosphodiester bond at its 5' carbon, as illustrated below

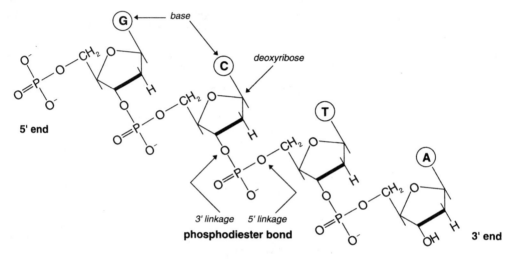

   b. When DNA is linear, usually there is a free 5'-phosphate group at one terminus and a free 3'-OH group at the other

C. DNA double helix

1. B-DNA, the principal form of DNA found in nature, consists of two strands of DNA wound around a common axis in a right-handed (direction of the twist) double helix (see *Right-Handed Double Helix of DNA*)
   a. Deoxyribosyl residues linked together by phosphodiester bonds form the helix backbone
   b. In each strand, the bases are at an angle to the helix backbone and project into the interior of the double helix; the base pairs lie in a plane that is close to 90 degrees to the helical axis
   c. The bases have **complementarity;** every base in one strand of the double helix is hydrogen-bonded with only its complement in the opposite strand
      (1) Adenine pairs only with thymine via two hydrogen bonds
      (2) Guanine pairs only with cytosine via three hydrogen bonds

Right-Handed Double Helix of DNA

In the right-handed double helix of DNA, the bases are complementary; every base in one strand of the double helix is hydrogen-bonded only with its complement in the opposite strand. As shown in the diagram below, adenine (A) always pairs with thymine (T), and guanine (G) always pairs with cytosine (C).

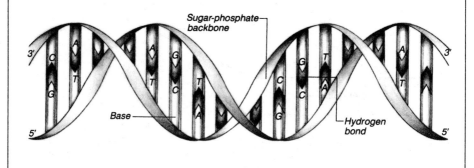

    d. The two single DNA strands are **antiparallel;** antiparallelism describes the relationship of the two DNA single strands to each other; one chain is oriented to begin at the 3'-OH group and end at a 5'-phosphate group (3' → 5'); its matching strand is oriented to begin at a 5'-phosphate group and end at a 3'-OH group (5' → 3')
2. Two other DNA forms, A-DNA and Z-DNA, occur less frequently
    a. A-DNA, the dehydrated form of B-DNA, is similar to B-DNA in many respects except that it forms a shorter, wider helix and its base pairs are tilted 20 degrees with respect to the helical axis; A-DNA is important when DNA binds to RNA
    b. Z-DNA, whose biological function is uncertain, is a left-handed double helix

D. Comparison of RNA and DNA
1. RNA resembles DNA in that it is composed of a sequence of nucleoside monophosphates joined through phosphodiester bonds; however, RNA differs structurally from DNA in several ways
    a. The sugar of RNA is β-D-ribose, instead of the β-D-deoxyribose of DNA
    b. RNA, like DNA, contains the purines adenine and guanine and the pyrimidine cytosine; however RNA contains the pyrimidine uracil rather than the pyrimidine thymine found in DNA
    c. RNA is a single-stranded, mostly linear structure, whereas DNA is a double-stranded helical structure
2. There are three major types of RNA: tRNA, mRNA, and rRNA; in contrast, there is only one major type of DNA (B-DNA)
3. RNA is located in both the nuclear region and cytosol of prokaryotes and in the nucleus and cytosol of eukaryotes; most DNA is located only in the nuclear region of prokaryotes or in the nucleus of eukaryotes

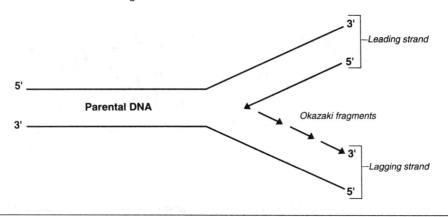

Replication Fork in DNA

The leading strand of the replication fork in DNA allows continuous replication in the 5' → 3' direction, starting from the 3' end of the template; the lagging strand, lacking a free 3' end, is replicated in the 5' → 3' direction in a discontinuous manner, thus producing small pieces of DNA called Okazaki fragments.

4. The function of all three major types of RNA is to direct and participate in various steps of protein synthesis (translation); the function of DNA is to transmit genetic information to daughter cells (replication) and to encode mRNA with the correct sequence of nitrogenous bases needed to synthesize proteins (transcription)

II. DNA Replication

A. General information
1. *Replication* is the process by which a single DNA double helix gives rise to two new DNA double helices with the same sequence of bases as in the original DNA; in replication, the cell synthesizes two new single DNA strands using each of the original pair of DNA strands as a template
   a. Replication takes place when cells divide to form daughter cells
   b. Replication allows the correct transmission of genetic information from the parent cell to daughter cells
2. Most information about DNA replication comes from studies of prokaryotes, which, having simpler intracellular organization, are easier to understand than eukaryotes
3. Replication of DNA is **semiconservative**; each new DNA molecule contains a single DNA strand from the original parent cell and one newly synthesized DNA strand
4. Replication occurs on both DNA strands simultaneously and in only the 5' → 3' direction
   a. Replication in the 5' → 3' direction cannot occur in the same way in both strands because DNA strands are antiparallel

b. The point at which the double-helical DNA unwinds and opens to allow replication is called the replication fork (see *Replication Fork in DNA*)
c. In the **leading strand** of DNA, which has a free 3' terminus, DNA synthesis starts at the 3' terminus and proceeds continuously in the 5' → 3' direction, catalyzed by the enzyme DNA polymerase III (DNA pol III)
d. In the **lagging strand** of DNA, which has a free 5' terminus, DNA synthesis starts at many points along the replication fork
    (1) The enzyme DNA pol III synthesizes small, discontinuous pieces of DNA, but only in the 5' → 3' direction
    (2) The small pieces of nascent DNA, called Okazaki fragments, are eventually joined by the enzyme DNA ligase

B. **Stages of prokaryotic DNA replication**
1. DNA replication occurs in four stages: prepriming, priming, elongation, and termination
2. DNA replication on the leading strand is continuous, beginning at the free 3' end; replication on the lagging strand results in discontinuous DNA segments that must be united into the final molecule
3. The *prepriming stage* prepares DNA for replication
    a. The enzyme *DNA gyrase* disrupts and relaxes the normal supercoiled structure (a coiled coil) of DNA
    b. Unwinding and stabilization of the double helix requires three proteins working in concert: helicase, rep protein, and single-strand binding (SSB) protein
        (1) Helicase moves along the lagging strand in the 5' → 3' direction; rep protein moves along the leading strand in the 3' → 5' direction; both enzymes act together to unwind and open the double helix just before the replication fork
        (2) SSB proteins bind to the separated single strands of DNA and prevent intrastrand base-pair formation (base-pairing within a single strand) (see *Replication of DNA on a Template Lagging Strand*, page 114)
4. In the *priming stage,* the enzyme *primase* synthesizes a special starting molecule called a **primer,** a short piece of RNA, hydrogen-bonded to the DNA template; the primer has a free 3'-OH group for attachment of new DNA
    a. Primase associates with several other priming proteins to form a complex called a *primosome*
    b. Primase does not require a primer for itself; it synthesizes the primer in the 5' → 3' direction without an attachment site
    c. Primase requires a DNA template, a piece of single-stranded DNA that directs the sequence of bases to be incorporated into the primer
5. The next main stages of DNA synthesis (elongation and termination) require two different enzymes—DNA polymerase I (DNA pol I) and DNA pol III
    a. *DNA pol III* is the active enzyme in elongation
    b. *DNA pol I* is the active enzyme in termination
    c. Another polymerase, DNA polymerase II, has also been isolated from prokaryotes, but its function is unknown

Replication of DNA on a Template Lagging Strand

The replication of DNA on the lagging strand is discontinuous. In the termination stage of replication, the various fragments of new DNA must be joined by the enzyme DNA ligase. In contrast to the replication mechanism illustrated below, replication on the leading strand of DNA is continuous.

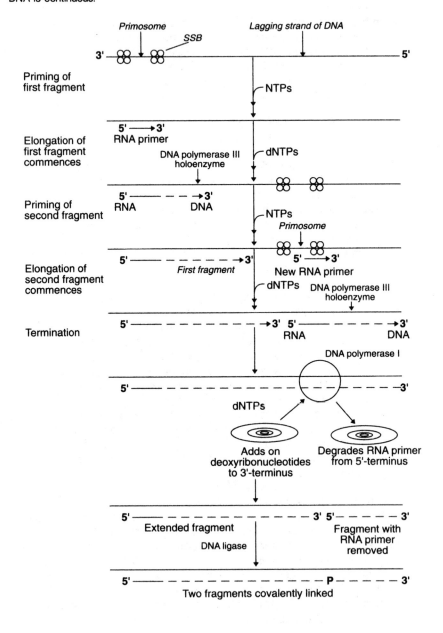

6. During the *elongation stage,* synthesis of new DNA takes place, beginning with the addition of the first deoxyribonucleotide to the free 3'-OH group of the RNA primer
   a. DNA pol III catalyzes DNA synthesis in the 5' → 3' direction only; first it adds a deoxyribonucleotide to the RNA primer followed by other deoxyribonucleotides to form new DNA
   b. Using a parental strand of old DNA as a template, DNA pol III catalyzes the sequential addition of deoxyribonucleotides to form new DNA
   c. Three components are required for DNA pol III activity: a DNA template, a pool of all four deoxyribonucleoside triphosphates (dATP, dGTP, dCTP, dTTP), and the cofactor magnesium (Mg^{2+}), which binds to the deoxynucleotide triphosphates
   d. DNA pol III also exhibits 3' → 5' **exonuclease** activity, a "proofreading" function that checks for and removes incorrect bases before DNA synthesis continues
7. The *termination stage* completes DNA replication
   a. As the new DNA strand elongates, DNA pol I removes the RNA primer and substitutes deoxyribonucleotides (a segment of new DNA) for the excised primer
   b. Besides removing the RNA primer, DNA pol I also can excise any incorrectly added bases and substitute the correct base; this is the end of the termination stage for DNA synthesis on the leading strand
   c. At the end of the termination stage on the lagging strand, *DNA ligase* links adjacent Okazaki fragments by forming a covalent bond between an adjacent 3'-OH and 5'-monophosphate group

C. Eukaryotic DNA replication
1. Although fundamentally similar to replication in prokaryotes, eukaryotic DNA replication is far more complex
   a. Eukaryotes have more DNA to replicate
   b. Eukaryotic DNA is extensively packaged to condense a large amount of DNA into the smallest possible space
2. Eukaryotic DNA is complexed with small basic proteins called **histones;** this complex is called **chromatin**
   a. Chromatin includes five different histone proteins classified as H1, H2A, H2B, H3, and H4
   b. In an electron micrograph, chromatin has the appearance of beads on a string
      (1) The "beads" are a complex of 146 base pairs (bp) of DNA wrapped around a histone core containing two molecules each of H2A, H2B, H3, and H4; this structure is called a **nucleosome**
      (2) The "string" that links adjacent nucleosomes is a complex of DNA and H1 plus other proteins
3. Polymerases that participate in eukaryotic DNA replication are not as well characterized as those in prokaryotic replication
   a. Polymerase α, found in the nucleus, is the major eukaryotic polymerizing enzyme, similar to prokaryotic DNA pol III; polymerase α also functions in eukaryotic DNA repair

b. Polymerase β, found in the nucleus, participates in DNA repair, similar to prokaryotic DNA pol I
   c. Polymerase γ, located in the mitochondria, replicates mitochondrial DNA

III. DNA Repair

A. General information
   1. Repair mechanisms ensure that errors in DNA base sequence or structure are not incorporated into new DNA during replication; if errors in the DNA base sequence or structure are not corrected, the error will be transmitted to progeny
   2. **Mutations,** errors in the DNA base sequence, can be classified into different types
      a. A substitution occurs when an incorrect base replaces the correct base
      b. A deletion occurs when one or more bases are missing
      c. An insertion occurs when one or more bases are added

B. Excision repair of pyrimidine dimers
   1. To repair pyrimidine dimers, cells use *excision repair,* one of the most well-characterized repair mechanisms
   2. A *dimer* is a covalent linkage between adjacent pyrimidine bases in the *same* DNA strand; dimers are produced in DNA as a result of cellular exposure to ultraviolet light; a dimer creates a structural distortion in the DNA
   3. DNA that contains pyrimidine dimers can neither replicate nor be transcribed correctly; this may have lethal consequences for the cell
   4. Excision repair of a dimer is a multistep process
      a. A specific DNA repair **endonuclease** recognizes the structural distortion and hydrolyzes the DNA strand containing the dimer; the endonuclease cleaves at a phosphodiester bond several bases from the dimer, leaving a gap in the DNA strand
      b. DNA pol I removes the DNA fragment that contains the error and catalyzes the addition of the correct nucleoside monophosphates at the 3′ terminus of the last nucleotide before the gap; the correct DNA strand serves as a template
      c. The enzyme DNA ligase joins the nascent DNA fragment to the rest of the strand

C. Other DNA repair mechanisms
   1. Photoreactivation, recombination repair, and SOS repair are other mechanisms to repair damaged DNA
   2. Photoreactivation repair uses the enzyme DNA photolyase, which, when activated by a certain wavelength of light, cleaves a pyrimidine dimer
   3. Recombination repair occurs in bacteria and involves exchange of nucleotides between two DNA double helices

a. DNA may replicate despite the presence of a lesion, such as a pyrimidine dimer; such a replication results in a pair of DNA double helices, with one double helix having perfect base-pair matching and the second double helix bearing a damaged parent strand and a daughter strand with a gap opposite the damaged area on the parent strand
b. Through action of the enzyme *RecA protein,* the needed DNA segment is removed from the parent strand of the perfect DNA and inserted into the nucleotide gap on the newly formed DNA daughter strand
c. The gap created on the parent strand is then filled with the appropriate nucleotides
4. In SOS repair, an excess of single-stranded DNA, which accumulates when DNA replication is inhibited, sends a distress ("SOS") signal to produce DNA repair enzymes; SOS repair occurs in the bacterium *Escherichia coli*

IV. DNA Transcription

A. General information
1. *Transcription* is the synthesis of RNA directed by a DNA template; all forms of RNA (mRNA, tRNA, and rRNA) are created through transcription
2. Only one strand of the double-stranded DNA is transcribed into RNA
   a. The DNA strand that is transcribed (called the template, coding, sense, or copy strand) is the $3' \rightarrow 5'$ strand
   b. The strand that is not transcribed (called the noncoding or antisense strand) is the $5' \rightarrow 3'$ strand
3. *RNA polymerase,* a large protein with five subunits, catalyzes prokaryotic DNA transcription
   a. The core enzyme consists of two α subunits, one β subunit, and one β' subunit
   b. The holoenzyme (see Chapter 4, Protein Function and Metabolism) consists of the core enzyme plus another subunit called σ, whose function is to recognize the **promoter** sequence of the DNA strand to be synthesized

B. Stages of transcription
1. Transcription occurs in three stages: initiation, elongation, and termination (see *DNA Transcription,* page 118)
2. DNA transcription occurs in the $3' \rightarrow 5'$ direction of DNA and produces a strand of $5' \rightarrow 3'$ strand of RNA
3. In the *initiation* stage, the RNA polymerase holoenzyme binds to double-stranded DNA at the promoter sequence, which is a section of DNA rich in thymine and adenine
   a. In prokaryotes, the promoter sequence is called the Pribnow box; in eukaryotes, the promoter sequence is called the Hogness box or TATA region because of the abundance of thymine (T) and adenine (A)
   b. The promoter is always located **upstream** from (before) the transcription start site
      (1) In prokaryotes, the promoter is about 10 base pairs (bp) upstream from the transcription start site

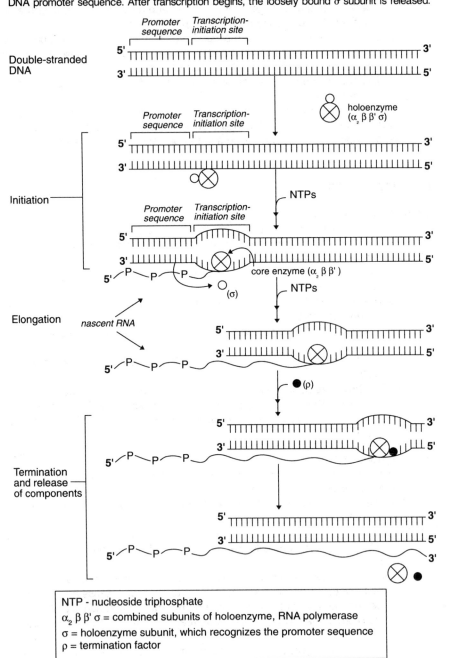

(2) In eukaryotes, the promoter is about 20 to 30 bp upstream from the transcription start site
4. After initiation, the σ subunit of the enzyme dissociates but leaves the RNA polymerase core enzyme bound to DNA while RNA synthesis continues; the continuous synthesis of RNA using the DNA template is called transcriptional *elongation*
5. In *termination,* nascent RNA is released from the RNA polymerase and the RNA polymerase core enzyme is released simultaneously from the DNA
   a. Simple termination occurs when the RNA polymerase encounters a DNA base sequence that contains a **palindromic sequence** (a base sequence that reads the same in both directions; it is also called an inverted repeat
      (1) In the DNA of *E. coli,* termination regions are rich in adenine (A) and thymine (T) and also contain a region rich in guanine (G) and cytidine (C) with a palindromic sequence just before the A-T-rich region
      (2) Transcription of these characteristic termination regions causes the RNA to form a "hairpin" or a "stem and loop" secondary structure that signals RNA polymerase to stop transcription
   b. Termination also may occur using a termination factor called ρ protein
      (1) ρ protein binds to cytosine-rich regions of RNA and signals the end of transcription
      (2) ρ protein is important when strong hairpin termination structures are not formed

C. **Comparison of eukaryotic and prokaryotic transcription**
   1. In eukaryotes, transcription uses three different RNA polymerases: RNA polymerase I (RNA pol I), RNA polymerase II (RNA pol II), and RNA polymerase III (RNA pol III)
      a. RNA pol I transcribes DNA that encodes rRNA
      b. RNA pol II transcribes DNA that encodes mRNA
      c. RNA pol III transcribes DNA that encodes tRNA and one rRNA
   2. In prokaryotic DNA, the entire nucleotide sequence is transcribed into RNA and translated into protein; in contrast, eukaryotic DNA has noncoding nucleotide sequences called **introns** (regions that do not encode functional RNA or protein) interspersed with fully coding nucleotide sequences called **exons** (regions that encode the exact base sequence found in RNA or protein)
      a. In eukaryotes, both the exons and introns of the DNA template are transcribed into RNA
      b. The RNA is then processed by small nuclear RNAs (snRNA) to remove (splice) the introns and link the exons, producing completely functional RNA
      c. Some introns catalyze their own excision; such self-splicing RNAs are called *ribozymes* (*ribo*nucleic acid en*zymes*)
   3. Eukaryotic mRNA has a 7-methylguanosine "cap" at its 5' terminus; the cap is important in mRNA splicing, stability, and translation into protein
   4. Eukaryotic mRNA has a polyadenylate (poly-A) "tail" at its 3' terminus; this poly-A region protects the mRNA from degradation and facilitates transport of mRNA from the nucleus to the cytoplasm, where translation will occur

D. The genetic code
1. A product of DNA transcription, mRNA contains a specific sequence of bases in the form of nucleoside monophosphates linked by phosphodiester bonds
2. The base sequence in mRNA directs the sequence of amino acids to be assembled as a protein at the ribosome; each sequence of three bases in mRNA, called a **codon,** calls for a particular amino acid to be incorporated into the growing polypeptide chain
3. The **genetic code** is the sequence of codons in mRNA that determines the sequence of amino acids in a protein
4. The genetic code has five unique characteristics; it is *triplet, degenerate, nonoverlapping, commaless,* and exhibits *"wobble"*
   a. The code is triplet because one codon (set of three bases) specifies one amino acid
   b. The code is degenerate because the four RNA bases (U, C, A, G) can be combined into a total of 64 different combinations ($4^3 = 64$); because there are only 20 amino acids to encode, more than one codon can code for the same amino acid
   c. The genetic code is nonoverlapping because no bases are shared between consecutive codons; each base in a sequence of bases is part of only one codon
   d. The genetic code is commaless; every base is part of a codon and there are no bases to indicate a pause in translation
   e. The genetic code exhibits "wobble" because the bases at positions 1 and 2 of the codon are specific for a particular amino acid, but the base at position 3 can vary (for example, GUA or GUG can encode the amino acid valine at position 3)
5. Of the 64 possible base combinations that can make up a codon, 61 code for amino acids
   a. The codon AUG codes for methionine; it is also the most common signal to initiate translation
   b. Three codons, UAG, UAA, and UGA, are signals to stop translation; they do not code for amino acids

V. RNA Translation

A. General information
1. *Translation* is the process by which the sequence of bases in mRNA directs the ribosomal synthesis of a protein; in translation, each codon of mRNA directs a particular amino acid to join the growing polypeptide
2. Translation occurs in the cytoplasm on either free ribosomes or ribosomes bound to the endoplasmic reticulum (ER)
3. Translation is a multistep process requiring many components
   a. Translation requires three different types of RNA: mRNA, tRNA, and rRNA
   b. Translation requires a supply of all 20 amino acids
   c. A group of enzymes, collectively known as aminoacyl tRNA synthetases, catalyze the linkage of an amino acid with its specific tRNA

d. Translation requires ribosomes, the organelles on which protein synthesis occur; ribosomes consist of two subunits that are masses of protein and rRNA
   (1) Prokaryotic ribosomes contain a 30S and a 50S subunit that join to form the functional 70S ribosome (the S stands for a particle's sedimentation coefficient in Svedberg units; sedimentation coefficients of a dissociated subunit are not additive; see Chapter 2, Protein Composition and Structure)
   (2) Eukaryotic ribosomes contain a 40S subunit and a 60S subunit that join to form the functional 80S ribosome
e. Translation requires an energy source, which is supplied through the hydrolysis of ATP and GTP; the hydrolysis of ATP and GTP requires the cofactor Mg^{2+}
f. Initiation factors, elongation factors, and release factors are proteins required for initiation, elongation, and termination of protein synthesis, respectively

B. **RNA's functions in translation**
 1. mRNA is the template that directs the sequence of amino acids incorporated into a protein; it is the "messenger" of the genetic code
    a. mRNA is synthesized in the nucleus of eukaryotes and in the nuclear region of prokaryotes
    b. The codons of mRNA are determined by the sequence of nucleotides on the template DNA
    c. Before translation begins in eukaryotes, mRNA must leave the nucleus, enter the cytosol, and attach itself to a ribosome, the site of protein synthesis
    d. Prokaryotic mRNA is polycistronic (one mRNA encodes more than one protein); eukaryotic mRNA is monocistronic (one mRNA encodes only one protein)
 2. tRNA molecules act as individual shuttles to transfer each amino acid to the correct site in the growing protein
    a. tRNA has an L-shaped structure containing three loops that resemble a cloverleaf; one of these loops is called the **anticodon** loop, which contains a sequence of three bases called the anticodon and base-pairs to its codon on the mRNA
       (1) Matching of anticodon to codon ensures that the correct amino acid, as directed by the mRNA template, is placed into the correct sequence in the growing protein
       (2) The anticodon-to-codon binding takes place through normal base pair hydrogen bonding
    b. Cells have many different tRNA molecules, at least one for each of the 20 amino acids
       (1) Each tRNA carries a specific amino acid, called its cognate amino acid
       (2) tRNAs are named according to the amino acid they transfer; for example, the tRNA that transfers glycine is designated $tRNA^{Gly}$ when glycine is not attached; $Gly\text{-}tRNA^{Gly}$ indicates the tRNA specific for glycine

c. The last three bases at the 3' terminus of all tRNA are 5'-CCA-3'; each cognate amino acid binds at the 3' terminal adenylate residue
d. A class of enzymes called aminoacyl-tRNA synthetases catalyzes the joining of the cognate amino acid with its specific tRNA
e. tRNA contains many unusual bases, such as methylated or reduced forms of adenine, cytosine, guanine, or uracil

3. rRNA is part of the ribosome structure and probably serves as the catalyst for the formation of bonds between individual amino acids in a protein
   a. In prokaryotes, 16S rRNA constitutes part of the small ribosomal (30S) subunit, whereas 5S and 23S rRNAs constitute part of the large ribosomal (50S) subunit
   b. In eukaryotes, 18S rRNA constitutes part of the small ribosomal (40S) subunit, whereas 5S, 5.8S, and 28S rRNAs constitute part of the large ribosomal (60S) subunit

C. Stages of translation

1. Translation in prokaryotes occurs in four stages: activation, initiation, elongation, and termination
2. In the *activation stage,* an amino acid bonds with its particular tRNA
   a. In the first step of activation, the carboxyl group of an amino acid bonds to the phosphoryl group of AMP to form an "activated" amino acid called aminoacyl-AMP
   b. AMP is derived from the hydrolysis of ATP to AMP and pyrophosphate; the energy to drive the activation reaction comes from the hydrolysis of pyrophosphate to two molecules of inorganic phosphate
   c. In the second step of activation, the aminoacyl group of aminoacyl-AMP is transferred to the tRNA specific for that amino acid
   d. The aminoacyl-AMP forms an ester bond with either the 2'- or 3'-OH in the 3' terminal adenine of its specific tRNA; the reaction produces an aminoacyl-tRNA and liberates AMP
   e. Both the activation and transfer steps are catalyzed by the same enzyme, aminoacyl-tRNA synthetase, which occurs in many specific forms
      (1) Each aminoacyl-tRNA synthetase has a high degree of specificity for a certain tRNA and amino acid; this enzyme specificity prevents incorporation of an incorrect amino acid into a protein
      (2) Aminoacyl-tRNA synthetase also checks and corrects the activation of an incorrect amino acid before it is attached to the tRNA
3. In the *initiation stage,* the functional 70S ribosome is assembled with the first aminoacyl-tRNA bound to mRNA at the site that encodes the start of protein synthesis (see *Initiation Complex in Prokaryotic mRNA Translation*)
   a. AUG, the initiation codon, codes for the amino acid methionine (Met); actually, N-formyl methionine, a formylated methionine that is the first amino acid in nascent protein of all prokaryotes, is brought to the AUG codon by a special t-RNA molecule, tRNA^{fMet}
      (1) N-formyl methionine is simply methionine with a formyl group ($-HC=O$) linked to the nitrogen of its α-amino group; the superscript *fMet* on tRNA^{fMet} stands for N-formyl methionine and indicates that this tRNA transfers N-formal methionine only

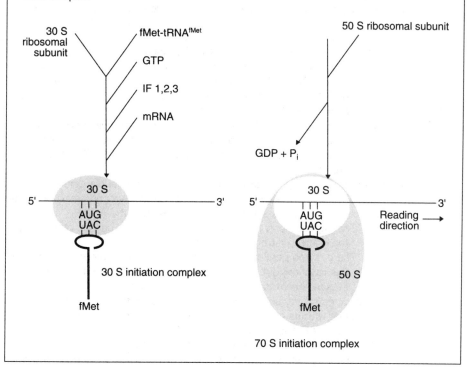

(2) tRNA^{fMet} reacts with methionine to form a Met-tRNA^{fMet} intermediate; the enzyme formyl transferase then adds a formyl group to methionine to form the initiating aminoacyl molecule, fMet-tRNA^{fMet}
(3) fMet-tRNA^{fMet} carries N-formyl methionine to the ribosome
b. The fMet-tRNA^{fMet} molecule combines with other components to form the 30S *initiation complex*
   (1) Prokaryotic ribosomes contain a 30S and a 50S subunit that join to form the functional 70S ribosome
   (2) The 30S initiation complex contains mRNA, fMet-tRNA^{fMet}, the 30S ribosomal subunit, and three initiation factors—IF1, IF2, and IF3; GTP is also required for formation of this complex
   (3) The anticodon region of fMet-tRNA^{fMet}, containing the base sequence 3'-UAC-5', base-pairs with the 5'-AUG-3' initiation codon in the mRNA
   (4) Codon-anticodon binding is facilitated by the 30S ribosomal subunit, whose 16S rRNA binds to a purine-rich sequence in the mRNA known as the Shine-Dalgarno sequence; this binding ensures that all components are properly aligned to begin protein synthesis

c. The 50S ribosomal subunit joins the 30S initiation complex to form the functional 70S initiation complex; hydrolysis of GTP provides energy for this process
4. During the *elongation stage,* amino acids, which are carried by their specific tRNAs and directed by the mRNA codons, are added sequentially to form a protein; elongation occurs in three steps: aminoacyl-tRNA binding, peptide bond formation, and translocation (see *Elongation Stage of Prokaryotic mRNA Translation*)
   a. At the start of Step 1, the fMet-tRNA^{fMet} is bound to a site on the 50S ribosomal subunit called the P (peptidyl) site
      (1) The A (aminoacyl) site of the ribosome contains the next unmatched mRNA codon; in the illustration, the unpaired codon GUU calls for a Valine-tRNA^{Val}
      (2) In Step 1, *aminoacyl-tRNA binding,* the correct aminoacyl-tRNA enters the A site and matches its anticodon with the mRNA codon; hydrogen bonds form between the bases of codon and anticodon
   b. In Step 2, *peptide bond formation,* the amino acid fMet dissociates from fMet-tRNA^{fMet} and forms a peptide bond with the second amino acid; the second amino acid is still attached to its tRNA at the A site
      (1) The peptide bond forms between the carboxyl carbon of fMet and the α-amino nitrogen of the incoming amino acid
      (2) The enzyme peptidyl transferase was previously thought to catalyze this reaction, but recent evidence suggests that rRNA, not the enzyme, facilitates peptide bond formation
   c. In Step 3, called *translocation,* the polypeptide (in this early stage, only a dipeptide) with its one attached tRNA moves from the A site to the P site on the 50S ribosomal subunit
      (1) The tRNA^{fMet} is released from the AUG codon
      (2) The entire ribosome, with P-site-bound tRNA and dipeptide, moves down one codon on the mRNA molecule with the second codon now aligned at the P site
      (3) The ribosome is now in place to translate the third codon by bringing in the correct aminoacyl-tRNA to the A site
   d. The process continues until all of the amino acids directed by mRNA codons and carried by tRNA are joined by peptide bonds
   e. Elongation protein factors EF-Tu, EF-Ts, and EF-G are required to position and check the accuracy of the tRNA in the ribosome
5. The *termination stage* marks the end of protein synthesis, resulting in the release of the nascent protein and all the synthetic components
   a. The signal to end translation is the appearance of a termination codon (UAG, UAA, or UGA) at the A site
   b. Protein release factors (RF1 and RF2) recognize a termination codon and cause hydrolysis of the nascent protein at the carboxyl terminus of the last amino acid from its tRNA
   c. Release factors, mRNA, deacylated tRNA, and the nascent protein are released and the 30S and 50S subunits dissociate; all components are then available for another cycle of translation

Elongation Stage of Prokaryotic RNA Translation

The elongation stage of translation produces a polypeptide chain from the mRNA codon sequence. Elongation occurs in three steps: Step 1 shows aminoacyl-tRNA binding at the A site on the ribosome. Step 2 shows peptide bond formation between adjacent amino acids. Step 3 shows translocation, the movement of the ribosome to the next mRNA codon and the transfer of the tRNA-polypeptide chain from the ribosomal A site to the P site.

Step 1.

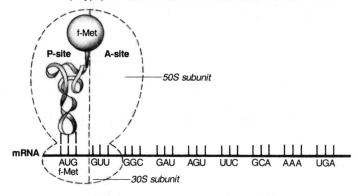

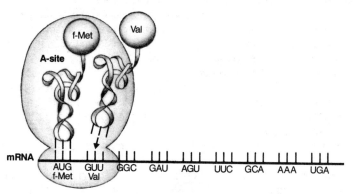

Step 2.

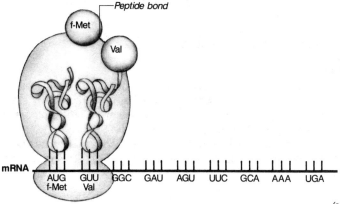

(continued)

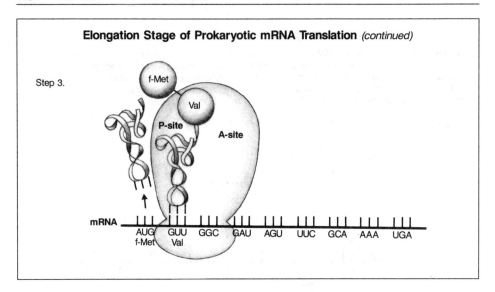

Elongation Stage of Prokaryotic mRNA Translation *(continued)*

Step 3.

D. **Comparison of eukaryotic and prokaryotic translation**
   1. Eukaryotes translate mRNA codons into protein in the same general manner as prokaryotes, but eukaryotes differ from prokaryotes in several respects
   2. Unmodified methionine is the amino acid used for initiation in eukaryotes; prokaryotes use N-formyl methionine
   3. Eukaryotic mRNA has no purine-rich Shine-Dalgarno sequence to align the correct AUG at the initiation site; instead, a eukaryotic 40S ribosomal subunit scans the mRNA by moving in the 5' → 3' direction until it encounters an AUG codon; translation begins at the first AUG codon the 40S subunit ribosome encounters
   4. The initiation, elongation, and termination factors of eukaryotes differ in composition, number, and complexity from those found in prokaryotes
   5. Prokaryotic mRNA is polycistronic, meaning that one mRNA encodes more than one protein; eukaryotic mRNA is monocistronic, meaning that each mRNA molecule encodes only one protein

VI. Post-Translational Processing

A. **General information**
   1. Cells modify proteins to direct them from their site of synthesis (free or bound ribosomes) to their appropriate intracellular or extracellular location
   2. Proteins used in the same cell in which they are synthesized are translated on free ribosomes (ribosomes not bound to the ER)
   3. Proteins exported from the cell are synthesized on ribosomes bound to ER (also called rough endoplasmic reticulum [RER]); ER with no ribosomes is called smooth endoplasmic reticulum (SER)

B. **Protein signaling for export**
   1. Proteins bound for export from the cell must enter the lumen of the RER, then undergo transport to the Golgi apparatus for further modification before leaving the cell
   2. As proteins translated on the RER are synthesized, they are targeted to the lumen of the RER by an amino acid **signal sequence** at their N-terminus
      a. The signal sequence is a group of 15 to 30 amino acids, many of which are hydrophobic
      b. The sequence recognizes the RER membrane by attaching to a transmembrane signal receptor
   3. The protein travels through the transmembrane signal receptor into the lumen of the RER, where enzymes called *signal peptidases* cleave the signal sequence from the protein

C. **Protein modification**
   1. Protein modification transforms a nonfunctional protein into a functional protein or chemically alters a protein so that it can participate in specialized reactions
   2. Proteins are modified in the lumen of the ER or in the Golgi apparatus
   3. Carbohydrates may be added to form glycoproteins in the process of glycosylation
      a. The type and number of carbohydrates that are added signal the protein's ultimate destination
      b. N-linked glycoproteins have carbohydrates attached to the amide group nitrogen of the amino acid asparagine
      c. O-linked glycoproteins have carbohydrates attached to the OH-group oxygen of the amino acids serine or threonine
   4. The formyl group of N-formyl-methionine and commonly one or more terminal amino acids are removed from the nascent protein
   5. Internal cleavage of amino acids may produce a functional protein from a nonfunctional precursor; an example is the processing of preproinsulin to proinsulin to the functional protein hormone insulin
   6. Chemical modifications of specific amino acids in specialized proteins may occur; an example is the addition of hydroxyl groups to proline residues to form hydroxyproline in collagen

Study Activities

1. Create a chart comparing and contrasting the composition, structure, and function of DNA and RNA.
2. Using a diagram of the structures of DNA, RNA, and protein, identify the major components of each type of molecule.
3. Explain the processes of replication, transcription, translation, and post-translational modifications in terms of cellular location, necessary components, and mechanisms involved; note how these processes differ in prokaryotes and eukaryotes.
4. Given the directional sequence of a piece of DNA and using the genetic code, decipher the sequence of bases in mRNA and the sequence of amino acids in a particular protein.
5. Define the terms listed in bold-faced type throughout this chapter.

8

Gene Regulation and Analysis

Objectives

After studying this chapter, the reader should be able to:
- Discuss positive and negative regulation of gene expression in prokaryotes, using the *lac* operon as an example.
- Discuss attenuation as a prokaryotic transcriptional control mechanism, using the *trp* operon as an example.
- Describe how oncogenes and tumor suppressor genes are involved in the malignant transformation of cells.
- Discuss the experimental techniques that may be used to isolate and identify a DNA segment encoding a particular protein

I. Regulation of Genetic Information

A. General information
1. A **gene** is a specific sequence of deoxyribonucleic acid (DNA) that encodes a sequence of messenger ribonucleic acid (mRNA) codons needed to synthesize a particular protein; a single DNA molecule contains a sequence of nucleotides corresponding to many different proteins and arranged as a sequence of genes on the DNA molecule
2. Genes are *expressed* as proteins only if they are transcribed into mRNA and then translated into protein; genes may be present in a cell but unexpressed because they are not transcribed into mRNA or translated into protein
3. In prokaryotes, regulation of gene expression occurs primarily at the level of transcription
4. In prokaryotes, the synthesis of enzymes needed to use a particular nutrient is controlled by a mechanism known as **enzyme induction;** this mechanism controls the synthesis of enzymes needed by the prokaryote *Escherichia coli* to use lactose
5. **Attenuation,** another mechanism affecting gene transcription and mrNA translation in prokaryotes, helps control the production of enzymes needed for tryptophan synthesis
6. The mechanisms for control of genetic transcription rely on various structures that occur in the gene sequence of DNA (see *Structure of an Operon*)

Structure of an Operon

A typical operon consists of adjacent genes working in a single coordinately controlled unit. Each of the structural genes in the *lac* operon (shown here as Z, Y, and A) encodes particular enzymes, which may or may not be synthesized, depending on whether a regulatory gene is bound at the control site. The promoter (P) and operator (O) control sites permit or repress the binding of RNA polymerase according to cellular needs for the enzymes. The regulatory gene I encodes a repressor protein, which in turn affects the activity of the control site genes. For some operons, the regulatory gene is on a nearby segment of DNA, as shown here; other operons are controlled by regulatory genes occurring some distance away on the DNA strand.

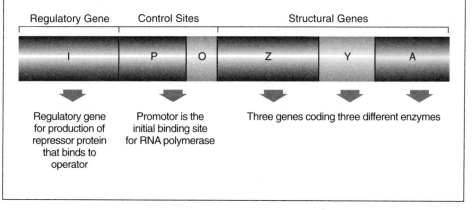

Regulatory Gene | Control Sites | Structural Genes

Regulatory gene for production of repressor protein that binds to operator

Promotor is the initial binding site for RNA polymerase

Three genes coding three different enzymes

    a. An *operon* is a DNA segment consisting of adjacent genes that are controlled and function in a coordinated manner; some operon genes may suppress expression of other genes in the operon; some operon genes may facilitate expression of other genes in the operon

    b. Structural genes are the sections of the operon that encodes the amino acid sequences needed to produce a particular protein or proteins; the structural gene encodes the mRNA that eventually is translated into a protein

    c. The *control sites* are those portions of the operon that control (enhance or repress) expression of the structural gene

    d. The *regulatory gene* encodes a repressor protein that represses expression of the structural genes by binding at a control site

7. **Oncogenes** are cellular genes that have lost normal regulatory mechanisms; oncogenes are always expressed and produce increased quantities of proteins that stimulate cell growth under normal conditions but may produce cancer in an uncontrolled state

8. Tumor suppressor genes are normal cellular genes that induce the production of growth suppressor proteins; cancer may occur when these genes are not expressed, resulting in insufficient quantities of proteins needed to suppress excessive cell growth

B. **Function and control of the *lac* operon**

    1. The *lac* operon is a segment of *E. coli* DNA that encodes the enzymes the bacterium needs to metabolize the sugar lactose

a. In *E. coli,* glucose is the preferred carbohydrate energy source; lactose is used only if glucose is unavailable
b. Genes that encode the enzymes required for lactose uptake and use are expressed only when glucose is unavailable
c. Use of lactose by *E. coli* is an example of enzyme induction—the enzymes needed for *E. coli* to use lactose are produced only under certain circumstances
d. The *lac* operon is an example of how a prokaryote coordinately regulates its ability to use lactose as an alternate energy source
2. The *lac* operon consists of structural genes and a control site, which determine if and how the cell can use lactose
   a. Three structural genes (Z, Y, and A) encode the three enzymes required for lactose uptake and use
      (1) Gene Z encodes β-galactosidase, which hydrolyzes the disaccharide lactose into the monosaccharides galactose and glucose
      (2) Gene Y encodes galactoside permease, which transports galactose and other sugars into cells
      (3) Gene A encodes transacetylase, whose function in lactose metabolism is unclear
   b. The control site has a **promoter** region, to which RNA polymerase binds, allowing the subsequent expression of the three structural genes; the control site also has an operator region, located upstream and near the structural genes
3. The regulatory gene (I) encodes a repressor molecule, a protein that prevents expression of the structural genes in the *lac* operon
   a. The repressor binds to the control site at the operator region and prevents transcription of the adjacent structural genes by blocking the attachment of RNA polymerase at the promoter region (see *The* Lac *Operon with Repressor Bound to Operator*)
   b. When the repressor is not bound to the operator, RNA polymerase is able to bind to the DNA at the promoter, allowing the structural genes (Z, Y, and A) to be transcribed
4. Enzyme induction in the *lac* operon involves two separate control paths: negative regulation and positive regulation
   a. *Negative regulation* refers to the repression of the structural gene when the repressor is bound to the operator
      (1) When glucose is available to *E. coli,* the repressor gene is transcribed and translated and its product, the repressor protein, binds to the operator; binding of repressor to operator prevents expression of the *lac* operon structural genes
      (2) When glucose is unavailable, the repressor gene is still transcribed and translated but another molecule (called an *inducer*) binds noncovalently to the repressor; the repressor-inducer complex cannot bind to the operator; negative regulation is overcome, and lactose uptake and use is initiated
      (3) When *E. coli* no longer requires lactose, the inducer dissociates from the repressor; the free repressor protein binds to the operator, and transcription of the structural genes is prevented once again

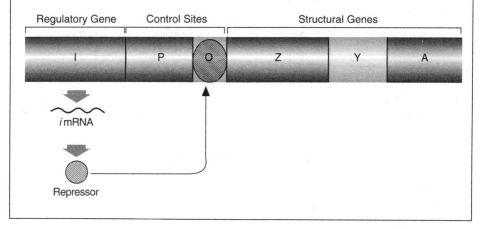

The *Lac* Operon with Repressor Bound to Operator

When the prokaryotic cell has sufficient glucose to meet energy needs, the enzymes for lactose metabolism are not needed for survival. In the glucose-rich state, the repressor molecule (indicated by the circle) binds to the operator section of the operon, thereby preventing the binding of RNA polymerase to the promoter and the synthesis of the structural genes, the *lac* operon enzymes.

    b. *Positive regulation* refers to expression of structural genes through the binding of a *positive regulator* to the promoter; for expression of its structural genes, the *lac* operon requires not only release from negative regulation (a free operator region), but also positive regulation
        (1) The positive regulator is adenosine 3',5'-monophosphate (also called cyclic AMP or cAMP)
        (2) cAMP binds to a dimeric protein called CAP (for *c*atabolite gene *a*ctivator *p*rotein or *c*yclic *A*MP-binding *p*rotein)
        (3) The cAMP-CAP complex binds at the *lac* operon promoter region
        (4) Binding of the cAMP-CAP complex is thought to facilitate separation of the DNA strands by RNA polymerase, allowing the structural genes to be expressed
    c. By itself, the *lac* promoter is weak; to induce expression of the structural genes, the promoter requires both positive regulation through cAMP-CAP binding plus release from negative inhibition through formation of the repressor-inducer complex (see *The* Lac *Operon Affected by cAMP-CAP Binding Plus Repressor-Inducer Complex,* page 132)

C. Function and control of the tryptophan (*trp*) operon
    1. In *E. coli,* the tryptophan (*trp*) operon controls the production of enzymes needed for synthesis of the amino acid tryptophan; other operons control the synthesis of other amino acids

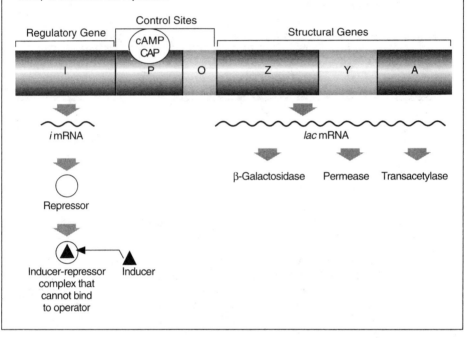

2. The *trp* operon is controlled by two different mechanisms: an operator-repressor mechanism, which somewhat resembles that found in the *lac* operon, and an *attenuator* mechanism, which relies on a unique base sequence (the *attenuator region*) at the beginning of the structural genes for the *trp* operon
3. The *trp* operon consists of structural genes and a control site similar to those of the *lac* operon (see *Structure of the* Trp *Operon*)
   a. The control site genes contain an operator and a promoter; these genes support or suppress transcription of the operon through the mediation of a repressor protein (*trp* repressor), which is the product of an independent operon called the *trp* R gene (not shown in diagram)
      (1) Repressor protein alone does not bind to the operator; expression of the entire *trp* operon continues as long as repressor protein is in its normal conformational state
      (2) Repressor protein binds to tryptophan to form a complex that binds strongly to the operator and reduces the transcription of the *trp* operon significantly
      (3) The repressor-tryptophan complex is called a *repressor-corepressor system* (tryptophan is the corepressor) because both molecules are required to stop transcription of the operon effectively

Structure of the *Trp* Operon

This diagram of the *trp* operon shows the control sites, which are subject to influence by a repressor-corepressor system, and the leader sequence, which controls synthesis of the structural genes E, D, C, B, and A through the mechanism of *attenuation*. The attenuator region encodes two tandem tryptophan codons, a unique feature that greatly influences attenuation of the operon.

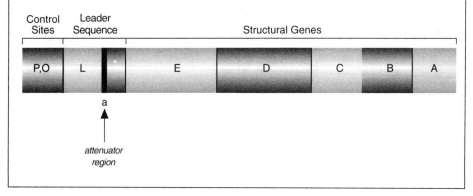

    b. The structural genes include genes E, D, C, B, and A plus a leader sequence (L)
       (1) Genes E, D, C, B, and A encode enzymes required by *E. coli* for tryptophan synthesis
       (2) The Leader sequence (L) consists of 162 nucleotides that precede E, the first structural gene
       (3) Within L (about 30 to 40 nucleotides upstream from the start of gene E) occurs the *attenuator region* (labeled a), a unique region encoding fourteen amino acids that includes a sequence encoding the base sequence UGG-UGG (tryptophan codons) in its corresponding mRNA; the ribosomal translation of the tandem pair of UGG codons calls for a pair of tRNAs bearing tryptophan molecules
4. *Attenuation,* the control of an operon through the action of the attenuator region, depends on three features: a unique sequence of nucleotides in the DNA of the attenuator region, the normal translation mechanism of prokaryotes, and the secondary structure of the mRNA transcribed from the attenuator region
    a. The DNA of the attenuator region encodes the unique tandem sequence UGG-UGG in its mRNA; reading the UGG-UGG codons in mRNA, the translating ribosome must incorporate two successive tryptophan molecules into the growing polypeptide chain; the availability of tryptophan affects how quickly the ribosome can move through the attenuator region mRNA
    b. In normal prokaryotic translation, the ribosome begins translation of mRNA before transcription is complete; because prokaryotes lack defined nuclei, mRNA is available for translation as it is being synthesized

c. The secondary structure of the attenuator region mRNA determines whether the structural genes will be transcribed and translated; the mRNA from this region can form two different secondary structures; both are hairpin structures that result from intrastrand base-pairing

5. When the intracellular tryptophan concentration is low, the attenuator region forms a hairpin structure between segments 2 and 3 (2-3 hairpin) and transcription of the entire *trp* operon proceeds (see the diagram below)

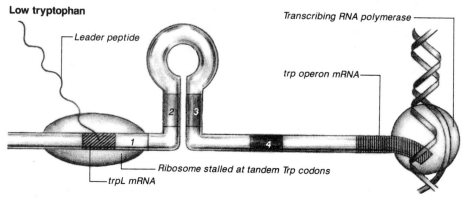

a. Low intracellular tryptophan results in a low concentration of tryptophan tRNA^{Trp} available for translation
b. The translating ribosome becomes "stalled" at the tandem UGG-UGG codons (the tryptophan codons) in the attenuator region of the leader sequence
c. The stalling of the ribosomes is a regulatory signal, causing segments 2 and 3 of the attenuator region to form a 2-3 hairpin; this signals RNA polymerase to continue transcribing the entire *trp* operon, including the structural genes encoding the enzymes needed for tryptophan synthesis

6. When the intracellular tryptophan concentration is high, the attenuator region forms a 3-4 hairpin and transcription of the entire *trp* operon stops (see the diagram below)

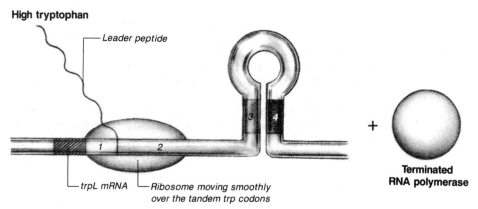

a. High intracellular tryptophan results in a high concentration of tryptophan tRNA^{Trp} available for translation
b. The translating ribosome are not stalled at the tandem UGG-UGG codons (the tryptophan codons) in the attenuator region of the leader sequence
c. The moving ribosome prevents bonding between segments 2 and 3 of the attenuator region; instead, intramolecular hydrogen-bonded base pairs form between segments 3 and 4 (3-4 hairpin)
d. The 3-4 hairpin is a termination signal for RNA polymerase; transcription is terminated at the end of the leader sequence and the structural genes are not transcribed

D. Uncontrolled gene expression in cancer
1. Under normal circumstances, controlled gene expression leads to controlled cell proliferation to replace dead or defective cells
2. Normal cell proliferation is mediated by proteins called *growth factors*
   a. Growth factors attach to cell membranes by binding to *growth factor receptors,* special transmembrane proteins
   b. The binding of growth factors to their receptors induces many intracellular changes (for example, activates phosphorylation of tyrosine residues on intracellular proteins); the growth factor receptor, upon binding with growth factor, acts as an intracellular tyrosine-specific kinase (an enzyme that catalyzes the addition of phosphate groups on proteins)
   c. Intracellular proteins with phosphorylated tyrosine residues may act as intracellular messengers that stimulate cell division
3. Normal cell proliferation also is mediated by hormones
   a. Binding of the hormone to its receptor causes a guanyl nucleotide binding protein (G-protein) to release guanosine diphosphate (GDP) and take up guanosine triphosphate (GTP); the G-protein–GTP complex in turn activates the enzyme adenylate cyclase
   b. Adenylate cyclase catalyzes the conversion of adenosine triphosphate (ATP) to cAMP; a rise in the concentration of cAMP correlates with cell division in normal cells
4. Abnormal proliferation of cells, a characteristic of cancer, is associated with oncogenes
   a. Oncogenes, originally discovered in certain viruses, are cellular genes that have lost their regulatory controls
   b. Proto-oncogenes are the normal cellular analogs of oncogenes and exhibit normal cellular behavior; they have the potential to become oncogenes when the cell is insulted or injured by various mechanisms (such as radiation, ingestion of carcinogens, or viral infection)
   c. Cancerous cells exhibit two characteristics that oncogenes promote: *cell immortality* (abnormally long cell life span) and *uncontrolled cell division*
   d. Oncogenes encode various proteins that accelerate cell proliferation; these abnormal proteins resemble normal intracellular proteins that mediate cell growth
      (1) Some oncogenes encode tyrosine kinases; phosphorylation of tyrosine residues in receptor proteins serves as a false signal that the cell has been stimulated by growth factor

- (2) Some oncogenes encode growth factors that resemble normal growth factors in their capacity to stimulate cell division; abnormal cell proliferation results from abnormally high concentration of growth factors
- (3) Some oncogenes encode growth factor receptors; a proliferation of growth factor receptors renders a cell abnormally responsive to circulating growth factors
- (4) Some oncogenes encode G proteins; G protein proliferation activates adenylate cyclase which in turn fosters an abnormal increase in intracellular cAMP, a signal for cell division
- (5) Some oncogenes encode nuclear proteins that bind to nuclear components directly; these proteins are thought to alter transcription
5. Over 50 oncogenes have been discovered; they are identified by a three-letter code derived from the associated tumor or from the tumor plus the species of origin; for example, *src* is an oncogene from the avian *sarc*oma virus, while *sis* is an oncogene from a *si*mian *s*arcoma virus
6. **Tumor suppressor genes** produce proteins that suppress cell growth; they stop the action of oncogenes
    a. Oncogenes are proto-oncogenes that have lost their control mechanisms and thus are always "on" (expressed)
    b. Tumor suppressor genes also can lose their control mechanisms and thus always "off" (unexpressed)
    c. It is likely that regulation of *both* oncogenes and tumor suppressor genes must be lost before cancer develops; cancer results only when oncogenes are "turned on" and tumor suppressor genes are "turned off"

II. Techniques for Analyzing DNA

A. General information
1. Scientists face many obstacles when working with DNA
    a. The DNA molecule is too large and complex to work with in its entirety; it must be cleaved in a controlled manner to produce small, manageable-sized fragments
    b. Eukaryotic DNA contains both functional regions *(exons)* and nonfunctional regions *(introns)* that must be separated from each other; the exons must be "tagged" in some way so that their presence may be detected
    c. The DNA of interest may be present in only small quantities
2. Recombinant DNA technology (also called molecular cloning and genetic engineering) combines a number of experimental techniques to manipulate and investigate the structure and function of DNA; this technology has enabled investigators to perform detailed analysis of genetic information

B. DNA fragmentation by restriction endonucleases
1. **Restriction endonucleases** are naturally occurring enzymes in prokaryotes

> **Two DNA Restriction Fragments**
>
> The diagram below shows the single-stranded ends that result from DNA cleavage by restriction endonuclease $EcoR_1$. The double-headed arrow marks the point of symmetry in the palindrome. The single-headed arrows mark $EcoR_1$ cleavage sites.
>
> $$5' - G - A - A \downarrow T - T - C - 3'$$
> $$3' - C - T - T \uparrow A - A - G - 5'$$

   a. These enzymes cleave DNA molecules within certain characteristic base-pair sequences; they protect an organism against viral invaders by cleaving and degrading any foreign DNA
   b. The cell's own DNA is spared because some of its bases at favored cleavage sites are methylated; the methyl groups signal the enzyme not to cleave its own DNA
   c. Investigators take advantage of the unique properties of restriction endonucleases and use them experimentally to cleave other prokaryotic and eukaryotic DNA
 2. Restriction endonucleases recognize and hydrolyze phosphodiester bonds between nucleotides at unique sequences called *palindromes* (see Chapter 7, Transmission of Genetic Information); restriction endonucleases cleave both strands of DNA at or near the point of symmetry in a palindrome
 3. Restriction endonucleases are abbreviated by three letters, denoting their bacterial origin, sometimes followed by further designations regarding the particular strain; for example, the restriction endonuclease $EcoR_1$ is obtained from E. coli strain *RY*13, type *1*

C. DNA mapping
 1. Experimental use of restriction endonucleases permits construction of a map or diagram showing the base sequences of an actual DNA molecule; knowledge of the DNA base sequences helps determine the location of mutations and is useful in the study of genetic diseases
 2. Construction of a DNA map is a multistep procedure
   a. Identical solutions of DNA are treated separately with several different restriction endonucleases, with each endonuclease cleaving at a specific site; each treatment yields DNA restriction fragments with known cleavage sites (see *Two DNA Restriction Fragments*)
   b. The DNA fragments from each treatment are then separated by electrophoresis on an agarose gel, a technique that separates particles on the basis of their charge-to-mass ratio (see Chapter 2, Protein Composition and Structure); the separated fragments are identified by one of two methods
     (1) The gel is stained with ethidium bromide, a fluorescent dye that binds to double-stranded DNA

(2) A fragment containing a 5'-terminus can be visualized by ***autoradiography*** if an isotope of phosphorus [^{32}P] was added to the 5'-phosphate group of DNA before the addition of restriction endonucleases
  c. By comparing the cleavage patterns from each of the different restriction endonuclease treatments, the order of the separated fragments may be determined; to determine the original sequence of bases, overlapping fragments are pieced together, much like solving a puzzle
D. Identification of specific DNA sequences
  1. A technique known as a **Southern blot** (named for its inventor, Edwin Southern) allows separation and identification of a DNA segment with a specific base sequence by transferring the DNA from the gel to nitrocellulose paper (if mRNA is also transferred, the technique is called a Northern blot; a Western blot refers to the transfer of protein)
     a. The DNA of interest is treated with restriction endonucleases to obtain distinct DNA fragments of various sizes
     b. The fragments are separated by electrophoresis on an agarose gel
     c. The gel containing the separated fragments is treated with a strong base, which converts all the DNA to its single-stranded form
     d. The gel is then covered with nitrocellulose paper and compressed with a heavy plate—a procedure that "blots" the gel fluid and its DNA through the paper; nitrocellulose has a very strong affinity for single-stranded, but not double-stranded, DNA
     e. The nitrocellulose paper blotted in this manner binds single-stranded DNA in the same position that it occupied on the original electrophoresis gel plate
  2. A DNA ***probe*** identifies the single strands of DNA now bound to the nitrocellulose paper
     a. The probe is a single-stranded DNA fragment (or RNA) with a defined base sequence that is complementary to, and hence will hydrogen-bond with, the DNA fragment of interest
     b. The probe, which may be chemically synthesized, is "tagged" for ease of detection by addition of a large number of radioactive nucleotides, such as an isotope of phosphorous [^{32}P]
     c. The nitrocellulose containing the DNA with fragments is immersed for a designated period in a solution containing the probe; excess or unreacted probe is then removed by repeated washing of the nitrocellulose by the same solution without the probe
     d. Probe molecules that remain on the paper have bound to complementary DNA fragments; the location of probe molecules is determined by autoradiography
        (1) Photographic film, placed over the nitrocellulose for hours or days, becomes blackened because of radioactivity from the phosphorous isotope; the probe's radioactivity produces a characteristic band on the film

(2) The band's location defines the DNA fragment of interest and confirms that a particular DNA base sequence has been isolated and detected from the unknown DNA molecule

E. **DNA cloning**
1. An investigator may use **DNA cloning** to obtain almost unlimited quantities of DNA from a particular gene
    a. Cloning refers to a group of techniques that allows production of unlimited quantities of a gene needed for study or other purposes
    b. One form of cloning is accomplished by inserting the desired DNA into an autonomously replicating gene, usually a gene carried by a *plasmid* or bacteriophage
    c. A second form of cloning uses DNA polymerase to synthesize DNA through chemical means
2. The cloning of eukaryotic DNA begins with an mRNA template that synthesizes DNA
    a. Because DNA coding regions (exons) are separated by noncoding regions (introns), cloning the entire DNA sequence of a gene yields much useless information; starting with mRNA enables investigators to synthesize a DNA molecule composed of only exons
    b. mRNA is first treated with *reverse transcriptase,* an enzyme that transcribes RNA to DNA (the reverse of normal prokaryotic and eukaryotic transcription in which DNA is transcribed into RNA)
    c. Using mRNA as a template, reverse transcriptase synthesizes a DNA strand called *complementary DNA (cDNA)* because its base sequence is complementary to the starting mRNA
    d. The duplex of mRNA joined to cDNA is treated under defined conditions and with various enzymes to remove the mRNA and to replicate the single strand of DNA, producing *double-stranded cDNA (ds cDNA)*
3. The ds cDNA is inserted into a *vector,* a carrier for autonomously replicating DNA; the vector is commonly a bacteriophage (a bacterial virus) or a bacterial plasmid (an extra DNA molecule distinct from the bacterial chromosome that is still replicated by the cell)
    a. The vector incorporates the ds cDNA into its normal genetic apparatus; the combined vector DNA and ds cDNA is called **recombinant DNA**
    b. Because replication of a virus or plasmid is independent of replication of the bacterium, the recombinant DNA is amplified (copied many times) through replication of the vector
    c. Antibiotic-resistant genes in plasmids can help determine if ds cDNA has been incorporated into the vector
        (1) A particular restriction endonuclease, which allows the insertion of ds cDNA, cleaves the plasmid at several antibiotic-resistant gene sites
        (2) Although all sites contain the insert, the ds cDNA insert may inactivate one or more antibiotic-resistant genes at specific sites
        (3) The inactivation of a gene for resistance to a particular antibiotic now makes the cell susceptible to that antibiotic

(4) A cell's unchanged antibiotic resistance may indicate that ds cDNA was not incorporated into the vector
(5) A tagged DNA probe is used to screen the plasmid and to isolate the particular site of interest

4. A cloning technique that does not require the use of living cells is **polymerase chain reaction** (PCR)
   a. The DNA of interest is heated to break hydrogen bonds between complementary base sequences and to form single-stranded DNA
   b. Two DNA primers (short fragments of DNA complementary to each DNA strand at regions near the sequence of interest) are chemically synthesized; synthesis of the correct primers is possible only if part of the DNA sequence to be cloned is already mapped
   c. The primers are introduced to the single-stranded DNA, and the reacting solution is cooled; cooling causes the primers to bond to their complementary regions on the single-stranded DNA
   d. New DNA is synthesized by initially adding nucleotides to the primers and subsequently adding the nucleotides to the nascent DNA, directed by the base sequence in the DNA template
      (1) A pool of the four nucleotides required for DNA synthesis is added to the reaction mixture
      (2) The enzyme DNA polymerase is added to catalyze the addition of nucleotides
   e. When the newly replicated DNA is complete, the temperature is increased to facilitate separation of the newly replicated DNA strands, which initiates a second cycle of the same procedure
   f. Repetition of subsequent steps allows an exponential increase in DNA with each cycle

F. Genetic engineering
1. One aim of genetic engineering is to produce a modified version of a particular gene, which then can be reinserted into cells or organisms to study the effect of the modification
2. Recombinant DNA is a single DNA molecule containing nucleotide sequences from two different sources
   a. One common source is a bacterial **plasmid,** an extrachromosomal circular piece of DNA
   b. The other source of DNA can be either prokaryotic or eukaryotic and depends on the DNA of interest
   c. DNA from both sources is combined and treated with the same restriction endonuclease, thus generating the same fragments with similar "sticky" ends (ends with the same unpaired bases)
   d. Under controlled experimental conditions, the plasmid DNA combines with DNA from the other source and forms a recombinant DNA called *covalently closed circular DNA (cccDNA)* that contains DNA from both sources
   e. The recombinant DNA is inserted as a plasmid into a host bacterium, where it replicates

f. The investigator screens the host cell for the recombinant by looking for the presence or absence of the gene product (either mRNA or protein); for example, if the recombinant DNA contained the gene necessary for the production of human insulin, the appearance of human insulin in the host cells indicates that the recombinant gene has begun to reproduce

Study Activities

1. Explain how the *lac* operon functions in the presence and absence of glucose.
2. Explain why attenuation could not regulate gene expression in eukaryotic cells.
3. List the function of the proteins that oncogenes encode. Explain how these proteins may play a role in the transformation of normal cells to malignant cells.
4. Discuss the normal role of restriction endonucleases in prokaryotes, and explain their use in experimental DNA manipulation.
5. Define a DNA probe, and explain how it is used in analyzing DNA.
6. Define the terms listed in bold-faced type throughout this chapter.

Appendix

Selected References

Index

Glossary

Absolute configuration—the precise arrangement of substituents at a chiral center in a molecule; if configuration is based on the Cahn-Ingold-Prelog system, the isomers are designated as either *R* or *S*; if based on the Fischer projection system, the isomers are designated as D or L

Active site—the specific amino acid residues of an enzyme to which the substrate must bind for catalysis to occur

Allosteric interaction—molecular interaction at one site in a protein that affects the protein's interactions at another site

Amino acids—building blocks of protein featuring a control carbon bonded to an amine group, a carboxylic acid group, a hydrogen atom, and a unique side chain

Amphipathic molecules—a term describing molecules that contain both polar (hydrophilic) and nonpolar (hydrophobic) groups

Anion—molecule carrying a net negative charge

Antibody—specific protein produced in response to a foreign antigen

Anticodon—a sequence of three bases on tRNA complementary to its mRNA codon

Antigen—specific foreign compound in the body that provokes an immune response

Antiparallel—opposite orientation of the two strands that make up a single molecule of DNA; one DNA strand begins with the 5'-phosphate group and ends with the 3'-OH group; its partner strand begins with the 3'-OH group and ends with the 5'-phosphate group

Apoenzyme—protein portion of an enzyme without the cofactor(s) required for its catalytic activity

Apoprotein—the protein portion of a molecule without its prosthetic group

Attenuation—prokaryotic transcriptional control mechanism for operons that regulate amino acid biosynthesis; the secondary structure of the mRNA, as it is transcribed and translated, determines whether the structural genes for amino acid biosynthesis are expressed

Autoradiography—an indirect, experimental technique to detect the presence of a molecule by detecting an isotope that has been attached to it; after electrophoresis, X-ray film is placed over the gel; darkened areas of film indicate the isotope and the position of the fragment

Buffer—solution of a weak acid and its conjugate base that resists large changes in pH; the proton-donating capacity of the weak acid coupled with the proton-accepting capacity of the conjugate base maintains constant pH until the buffering capacity of the system is exceeded

Cation—molecule carrying a net positive charge

Chiral—stereoisomers that cannot be superimposed on their mirror images

Chromatin—complex of eukaryotic DNA and histones

Codon—sequence of three bases in mRNA that codes for a particular amino acid or a stop signal for translation

Coenzyme—organic molecules that serve as cofactors for certain enzymes

Cofactor—nonprotein component necessary for optimal enzyme activity

Commitment step—first irreversible enzyme-catalzyed reaction in a metabolic pathway that is unique to the pathway; the commitment step obligates the product of the reaction to continue through the pathway

Complementarity—ability of adenine to form a hydrogen bond only with thymine, and guanine to form a hydrogen bond only with cytosine in DNA

Covalent bond — strong bond where two atoms share an electron pair

Covalent modification — covalent attachment of an atom or molecule to another molecule; covalent modification usually changes a compound's activity level

Deamination — removal of the α-amine group from an amino acid

Denaturation — treatment with heat or chemical reagents, resulting in the loss of a protein's secondary, tertiary, and quaternary structures, and thereby rendering it inactive

Deoxyribonucleic acid (DNA) — a polymer of nucleic acids joined through phosphodiester linkages; the DNA molecule is a double helix composed of antiparallel single strands of DNA joined through hydrogen bonding of complementary bases

Diastereomers — stereoisomers that are not enantiomers

Dipole — asymmetric distribution of electrical charge on a molecule resulting in a partial positive charge in one region and a partial negative charge in another

Disaccharide — a sugar composed of two simple sugars (monosaccharides)

DNA cloning — production of identical copies of DNA through an autonomously replicating genetic element, usually a plasmid or virus

DNA probe — a single-stranded DNA fragment with a defined base sequence that is complementary to, and hence will hydrogen bond with, a DNA fragment of interest; the probe is "tagged" for ease of detection by addition of an isotope of phosphorous, ^{32}P

Enantiomers — a pair of molecules that are nonsuperimposable mirror images of each other; enantiomers are physically and chemically identical; they differ only in how they affect plane-polarized light; enantiomers are designated $(+)$ or $(-)$, depending upon whether a solution of the enantiomer rotates a plane of polarized light to the right $(+)$ or to the left $(-)$

Endergonic — energy-requiring reaction that has a positive $\Delta G^{0'}$ and will not proceed spontaneously

Endonuclease — enzyme that hydrolyzes internal phosphodiester bonds of nucleic acids

Energy charge — index of the energy status of a cell determined by comparing the relative intracellular proportions of ATP, ADP, and AMP, which ranges from 0 (no ATP) to 1 (all ATP); the formula for the index is

$$\frac{[ATP] + \tfrac{1}{2}[ADP]}{[ATP] + [ADP] + [AMP]}$$

the higher the proportion of ATP, the higher the cell's energy charge

Enzyme induction — the synthesis of enzymes needed to use a particular nutrient; the enzymes are synthesized only as the need for them arises

Eukaryotic cell — complex cell that contains a nucleus, membrane-bound organelles, and more than one chromosome

Exergonic — energy-releasing reaction that has a negative $\Delta G^{0'}$ and proceeds spontaneously (without the input of outside energy)

Exons — regions of eukaryotic DNA that encode protein

Exonuclease — enzyme that hydrolyzes either 3' or 5' terminal phosphodiester bonds of nucleic acids

Fatty acid — a hydrocarbon chain with a terminal carboxyl group

Fibrous protein — a water-insoluble protein with high tensile strength that usually contains interstrand cross-links among subunits; for example, collagen

Fischer projection formula — a schematic representation of a compound illustrating the orientation of bonds with respect to a chiral center; it is based upon the orientation of substituents at the chiral center in glyceraldehyde

Gene — a specific sequence of DNA that encodes a particular protein

Genetic code — sequence of codons in mRNA that specify the sequence of amino acids in a protein

Globular protein — a water-soluble protein, such as hemoglobin, that occurs in nature as a compact spheroidal molecule and that frequently contains quaternary structure

Glucogenic — referring to amino acids that are catabolized to pyruvate or tricarboxylic-acid-cycle intermediates so named because they are precursors of glucose

Gluconeogenesis — synthesis of glucose from noncarbohydrate precursors such as lactate, amino acids, and glycerol

Glycogenesis — synthesis of glycogen from glucose; a synthesis pathway that occurs mostly in liver and muscle cells to reduce high blood glucose

Glycogenolysis — successive removal of individual glucose molecules from glycogen (glycogen breakdown); a reaction that occurs in liver and muscle cells in response to low blood glucose levels

Glycolysis — set of enzyme-catalyzed reactions in which glucose is converted to pyruvate

Glycosylation — covalent addition of one or more carbohydrate residues to a protein, forming a glycoprotein

Histones — small basic proteins that bind to eukaryotic DNA)

Holoenzyme — a functional enzyme composed of an apoprotein with its cofactor(s) attached

Hormones — potent molecules (polypeptide, steroid, or amine) that are produced at one site in the body but travel to another site (their target) to exert their effects

Hydrogen bonds — forces that arise when a single hydrogen atom in a molecule bonds with other electronegative atoms, such as nitrogen or oxygen

Hydrophobic interactions — interactions that occur when nonpolar groups cluster together to exclude water

Introns — regions of eukaryotic DNA that do not encode protein; also called intervening sequences

Ionic bonds — bonding that occurs between a group that carries a positive charge and another group that carries a negative charge

Isoelectric point — the pH at which a molecule has no net charge and exists as a zwitterion; abbreviated as pI

Isoenzymes — enzymes with the same atomic structure but different atomic configurations, or enzymes composed of multiple subunits and arranged in various combinations; they catalyze the same reactions at different body sites or different stages of development

Isomers — compounds that have the same chemical formula but that differ in the arrangement of atoms

Kinase — an enzyme that catalzyes the addition of a phosphate group to a protein

Lagging strand — DNA strand that runs $3' \rightarrow 5'$ and is replicated in short, discontinuous fragments in the $5' \rightarrow 3'$ direction

Leading strand — DNA strand that runs $5' \rightarrow 3'$ and is replicated continuously in the $5' \rightarrow 3'$ direction

Messenger RNA (mRNA) — a molecule that forms from a DNA template through a process called transcription; mRNA codons are translated by ribosomes to form proteins

Micelles — aggregates of amphipathic molecules in an aqueous environment; micelle formation permits the polar groups to shield nonpolar groups from water molecules

Monosaccharide — a simple sugar, such as glucose, fructose, or galactose, to which no other sugar molecules are attached

Mutation — error in the base sequence of DNA which, if uncorrected, will be passed on to progeny

Nonpolar — having uniform electrical charge distribution and no dipole; electrons are equally shared; nonpolar molecules do not interact with water and are insoluble in aqueous solvents

Nucleoside — a molecule composed of a purine or pyrimidine bonded to a pentose sugar (ribose or deoxyribose) through a β-N-glycosidic bond

Nucleosome — a complex of 146 base pairs of eukaryotic DNA wrapped around a core consisting of 2 each of the histone proteins 2A, 2B, 3, and 4, and linked by histone H_1; this forms part of the overall structure of chromatin

Nucleotide — a nucleoside with one, two, or three phosphate groups attached to the hydroxyl group of the 5′ carbon

Oligosaccharide — any carbohydrate containing between two and eight monosaccharide units

Oncogenes — cellular genes that have lost their regulatory mechanisms; oncogenes encode cell growth proteins that may lead to cancer

Operon — DNA segment consisting of adjacent genes that are controlled and function as a coordinated unit

Oxidative phosphorylation — a series of mitochondrial reactions whereby the oxidation of NADH and $FADH_2$ in the respiratory chain is linked to the phosphorylation of ADP, generating energy in the form of ATP

Palindromic sequence — a sequence of DNA bases that reads the same on both complementary strands, provided both strands are read in the same direction, for example, in the 5′ to 3′ direction; also called an inverted repeat

Peptide — a small number of amino acids linked through peptide bonds

pH — negative logarithm, to the base 10, of the molar hydrogen ion concentration

Phosphodiester bond — two oxygen atoms of one phosphate group participating in two ester linkages with two other molecules

Phospholipids — a class of lipids in which a phosphate group attached at the *sn* 3 carbon and fatty acids attached at the *sn* 1 and *sn* 2 carbons of glycerol (forming glycerophospholipids) or sphingosine (forming sphingolipids)

Photosynthesis — process that occurs in chloroplasts of plants, in which light energy converts carbon dioxide and water to glucose and oxygen

Plasmid — autonomously replicating circular DNA found in some bacteria that has been added to the normal bacterial chromosome, hence sometimes called an accessory chromosome; it is used in recombinant DNA technology

Polar — having regions of electrical charge resulting from unequal sharing of electrons between atoms; polar molecules interact with water and are therefore soluble in aqueous solvents

Polymerase chain reaction — *in vitro* experimental technique that produces many copies of a DNA fragment when original quantities of DNA are limited

Polysaccharide — a carbohydrate of more than eight monosaccharide units; also called a complex sugar

Primer — a short piece of RNA hydrogen-bonded to a DNA template in the first stage of DNA replication; the primer has a free 3′-OH group for attachment of new DNA

Prokaryotic cell — a cell that lacks a nucleus and membrane-bound organelles and has only one chromosome which consists of double-stranded DNA, such as a bacterial cell

Promoter — the adenine-thymine rich region of DNA to which RNA polymerase binds in preparation for transcription

Propeptide — a polypeptide or protein that requires processing before it has functional activity

Prosthetic group — the nonprotein portion of a molecule that is necessary for the activity of the molecule

Protein — amino acid residues linked by peptide bonds to form polymers

Proton-motive force — the pH gradient and electrical potential across the mitochondrial membrane that couples the flow of electrons in the respiratory chain to the phosphorylation of ADP to yield ATP

Purine — a nitrogen-containing compound consisting of a six-membered ring fused with a five-membered ring; either adenine or guanine

Pyrimidine — a nitrogen-containing compound consisting of a six-membered ring; either cytosine, thymine, or uracil

Racemic mixture — mixture containing an equal amount of (+) and (−) enantiomers designated as (±)

Rate-limiting step — enzyme-catalyzed reaction in a metabolic pathway that controls the rate of subsequent reactions

Receptor — a specific protein component on a cell membrane to which a specific messenger molecule binds; binding elicits a series of events that cause a biochemical change in the cell

Recombinant DNA — DNA which has been experimentally altered to contain DNA from two different sources

Replication — a process that occurs whenever a cell divides to produce daughter cells; DNA duplicates itself so that daughter cells contain the same DNA molecules as the parent cell

Restriction endonucleases — prokaryotic enzymes used in recombinant DNA technology to recognize and cleave double-stranded DNA only at specific sites called palindromes

Semiconservative replication — the sole mode of DNA replication in which newly replicated DNA contains one parent strand and one synthesized strand

Signal sequence — a group of 15 to 30 N-terminal amino acids that directs a protein into the lumen of the endoplasmic reticulum during protein synthesis

Signal transducers — molecules that mediate the intracellular response to an extracellular signal (for example, the binding of a hormone to its receptor)

Southern blot — technique used to transfer DNA fragments to nitrocellulose paper after separation by electrophoresis

Stereoisomers — compounds which differ only in the arrangement of substituent groups around an atom

Template — a strand of DNA that directs the sequence of bases incorporated into a newly synthesized strand of DNA (in replication) or mRNA (in transcription)

Transamination — transfer of an α-amine group from an amino acid to an α-keto acid to yield a new amino acid and the keto acid of the original amino acid

Transcription — process of RNA synthesis directed by DNA base sequence

Translation — process by which ribosomes "read" the codons in an mRNA molecule and accept the tRNA bearing the correct amino acid needed for protein synthesis

Triacylglycerol — lipid composed of three fatty acids joined in an ester linkage to one of the carbon atoms of glycerol

Tricarboxylic acid cycle (TCA) — series of aerobic, enzyme-catalyzed reactions transforming acetyl CoA into CO_2

Tumor suppressor genes — genes that regulate cell growth so that cell division occurs only under defined conditions

Turnover number — rate at which a particular enzyme transforms a substrate into a product (moles of substrate transformed per minute per mole of enzyme under optimal conditions)

Upstream — moving along DNA in the $3' \rightarrow 5'$ direction

van der Waals forces — individually weak but collectively strong attractive or repulsive forces between optimally spaced molecules

Vector — DNA (from a plasmid or a virus) that "houses" a piece of foreign DNA; the vector, along with the DNA insert comprises recombinant DNA

Zwitterion — charged compound with a net charge of zero; the sums of positive and negative charges are equal

Selected References

Alberts, B., et al. *Molecular Biology of the Cell,* 2nd ed. New York: Garland Publishing, Inc., 1989.

Armstrong, F.B. *Biochemistry,* 3rd ed. New York: Oxford University Press, 1989.

Carey, F.A. *Organic Chemistry.* New York: McGraw-Hill Book Company, 1987.

Devlin, T.M. *Textbook of Biochemistry with Clinical Correlations,* 3rd ed. New York: John Wiley & Sons, 1992.

Rawn, J.D. *Biochemistry.* Burlington, N.C.: Neil Patterson Publishers, Carolina Biological Supply Company, 1989.

Stryer, L. *Biochemistry,* 3rd ed. New York: W.H. Freeman and Company, 1988.

Voet, D., and Voet, J.G. *Biochemistry.* New York: John Wiley & Sons, 1990.

Index

A
Absolute configuration, 4, 5i
Acetaldehyde, 2i
Acetic acid, 2i, 11i
Acetone, 2i
Acetyl coenzyme A, 38, 102
Acid
   amino, 21-27, 31-33, 71-75
   bile, 96
   definition of, 10
   deoxyribonucleic, 106
   equilibrium constant of, 11-12
   fatty, 93, 94, 99-104
   ribonucleic, 106
Acidosis, 15
Actin, 65-66
Active site, 59
Active transport, 98-99
Adenosine monophosphate (AMP), 3i
Adenosine triphosphate (ATP), 37, 52-53, 109i
Affinity chromatography, 34
Alanine, 2i, 24, 25
   titration curve for, 28
Alkalosis, 15
α helix, 30
Allosteric effect, 57, 63
Amido group, 2i
Amino acids
   anabolism of, 74-75
   catabolism of, 72-74
   chemical properties of, 25-26
   composition of, 31-32
   deamination of, 71
   definition of, 21
   essential, 74t
   ionization of, 26-27
   nonessential, 74t
   sequence of, 32
   structure of, 22i, 23t, 24-25
   transamination of, 71-72
Amine group, 22i
Anabolism, 36, 74-75
Anion, 34
Anticodon, 121
Antigens, 66-67
Apoenzyme, 59
Apoprotein, 66
Apparent dissociation constant, 12
Arginine, 25, 26
Aspartate, 24, 26
Attenuation, 128, 133i
Autoradiography, 138

B
Base, definition of, 10
β pleated sheet, 30
Bile acids, 96
Bioenergetics, 36-54
1,3, bis-phosphoglycerate, 3i
Bohr effect, 58

Bonds
   covalent, 8
   hydrogen, 7
   ionic, 7
   peptide, 21, 29
   thioester, 38
Buffer, 10, 14-18

C
Cahn-Ingold-Prelog system, 4
Cancer, oncogenes and, 135-136
Carbohydrates
   functions of, 76-81
   metabolism of, 81-92
   structure of, 76-81
Carbon, bonding properties of, 1-8
Carbonyl group, 2i
Carboxyl group, 2i, 22i
Catabolism, 36, 72-74
Cation, 25, 34
Cell
   eukaryotic, 19-20, 39
   prokaryotic, 18, 39
Cellulose, 81
Chemiosmotic hypothesis, 53
Chiral center, 3, 5i
Chiral stereoisomer, 3
Chloroplast, 20
Cholesterol, 95-96, 105
Chromatin, 115
Chromatography, 34
Citric acid cycle, 37, 45-46, 47i, 48
Cloning, 139
Codon, 120
Coenzymes, 59
Cofactors, 59
Collagen, 30, 63
Commitment step, 41, 88
Complementarity, 110
Cori cycle, 91
Covalent bond, 8
Covalent modification, 39, 85
Cysteine, 2i, 25-26
Cystine, 2i, 26
Cytochromes, 50-51
Cytoskeleton, 20
Cytosol, 18, 19

D
Deamination, 71
Denaturation, 33
Deoxyribonucleic acid (DNA), 106
   analysis of, 136-141
   cloning of, 139
   composition of, 107-110
   double helix of, 110, 111i
   mapping of, 137
   probe of, 138
   recombinant, 136, 139
   repair of, 116
   replication of, 112i, 113, 114i, 115
   RNA and, 111
   transcription of, 106, 117, 118i, 119-120

Diastereoisomer, 3
Diffusion, 98
Dimer, 116
Dipole, 7
Disaccharides, 80-81
Dissociation, 12, 57, 58i
Disulfide group, 2i

E
Edman degradation, 32
Electron transport chain, 48-49, 50i, 51-54
Electrophoresis, 34-35
Enantiomer, 3
Endergonic reaction, 37
Endonuclease, 116, 137
Endoplasmic reticulum, 19
Energy
   charge, 39, 72
   release of, 38-39
   sources of, 37-38
Enzymes, 59-60, 61i, 62-63
   induction and, 128
Equilibrium constant, 11-12
Ester group, 2i
Ethanol, 2i
Eukaryotes, 19-20, 39, 119, 126
Excision repair, 116
Exergonic reaction, 37
Exons, 119
Exonuclease, 115

F
Fatty acids, 93
   catabolism of, 99-102
   oxidation of, 100i
   structure of, 94i
   synthesis of, 102-103, 104i
Fermentation, 39
Ferritin, 66
Fischer projection system, 4, 6, 79i
Food, oxidation of, 38-39
Fructose, 41, 44, 79i
Functional groups, 2t

G
Gel filtration chromatography, 34
Genes, 128
   tumor suppression and, 136
Genetic code, 107, 120
Genetic engineering, 140-141
Gibbs free energy, 36
Glucokinase, 40
Gluconeogenesis, 82, 88, 89i, 90-92
Glucose, 40, 79i, 81, 83i, 84i
Glutamate, 24, 26
Glyceraldehyde, 4
Glycerol-phosphate shuttle, 52
Glycerophospholipids, 94
Glycine, 24, 25
Glycogenesis, 81, 84i, 85-86
Glycogenolysis, 82, 83i, 84, 85-86
Glycolipids, 95

i refers to an illustration; t, to a table.

Index

Glycolysis, 37, 42i-43i
  definition of, 39
  oxidoreduction-phosphorylation stage of, 41, 43-44
  priming stage of, 40-41
  splitting stage of, 41
Glycoproteins, 68, 69
Glycosylation, 69
Glyoxylate cycle, 48
Glyoxysome, 20
Golgi apparatus, 20

H

Haworth projection formula, 6
Helix, 110, 111i
Heme, 56
Hemoglobin, 56-57, 58i, 59
Henderson-Hasselbalch equation, 12i, 13i, 14-15
Hexokinase, 40
Histidine, 25
Histones, 70, 115
Holoenzyme, 59
Hormones, 19, 68
Hydrogen bond, 7
Hydrolase, 60
Hydroxyl group, 2i

I-J

Imidazole group, 25, 26
Immunoglobulins, 66-68
Inhibitors, 62, 63, 64i
Insulin, 68
Introns, 119
Ion exchange chromatography, 34
Ionic bond, 7
Ionization constant, 12
Iron protoporphyrin IX, 56i, 57
Isoelectric focusing, 35
Isoelectric point, 27
Isoenzyme, 40, 59
Isoleucine, 24, 25
Isomer, 24
Isomerase, 60

K

Ketone bodies, 73
Kinase, 38
Kinetics, 60-61
Krebs cycle, 37, 45-46, 47i, 48

L

Lactose, 80i
Leucine, 24, 25
Ligase, 60
Lineweaver-Burk plot, 62, 63i, 64i
Lipids
  functions of, 93
  metabolism of, 99-105
  structure of, 93-96
  types of, 93-96
Lipoproteins, 68, 69-70, 99
Lyase, 60
Lysine, 24
Lysosome, 20

M

Malate-aspartate shuttle, 52
Membrane, cellular, 97i, 98
Metabolism
  of carbohydrates, 81-92
  energy sources and, 36-38
  of lipids, 99-105
  oxidation and, 38-39
  of protein, 70-75
  regulation of, 39
Methionine, 24, 25
Micelle, 7, 8i
Michaelis-Menten equation, 60-61, 62i
Mitochondria, 19
Molecule
  amphipathic, 7, 8i
  hydrophobic, 7
Monomer, 27
Monosaccharides, 77-80
Muscle contraction, 65-66
Mutations, genetic, 116
Myofibril, 65
Myoglobin, 58i
Myosin, 65-66

N

Nitrogen fixation, 74
Nonpolar group, 7
Nucleoproteins, 68, 70
Nucleoside, 108
Nucleosome, 115
Nucleotide, 109i, 110

O

Oligosaccharides, 81
Oncogenes, 129, 135
Operon
  lac, 129i, 130, 131i, 132i
  tryptophan, 131-132, 133i, 134-135
Organelle, 18
Oxidation, 38-39, 100i
Oxidoreductase, 60
Oxygen dissociation curve, 57, 58i

P-Q

Palindromic sequence, 119
Passive transport, 98
Pentose phosphate pathway, 86, 87i, 88
Peptide, 21, 22i, 27, 29i
Peroxisome, 20
pH, measurement of, 10, 11i
Phenylalanine, 3i, 24, 25
Phosphatidyl choline, 2i
Phosphodiester, 107
Phospholipids, 93
Phosphorylation, oxidative, 39
  ATP generation and, 52-53
  description of, 48-49
  energetics of, 49-50
  respiratory chain and, 50-52, 54
Photoreactivation, 116
Photosynthesis, 20
Plasma membrane, 18, 19
Plasmid, 139
Polar group, 7

Polymer, 21
Polymerase chain reaction, 140
Polysaccharides, 81
Primer, 113
Prokaryotes, 18, 39, 119, 126
Proline, 24, 26
Propeptide, 65
Prosthetic group, 56
Proteins
  actin, 65-66
  collagen, 63-65
  composition of, 21-27
  conjugated, 68-70
  as enzymes, 59-63
  ferritin, 66
  fibrous, 63
  functions of, 55-70
  globular, 56
  hemoglobin, 56-57, 58i, 59
  hormones and, 68
  immunoglobulins, 66-68
  insulin, 68
  metabolism of, 70-75
  modification of (genetics), 127
  myosin, 65-66
  structure of, 27-31, 63
  study of, 31-35
Proton-motive force, 49, 52-53
Protoporphyrin, 56i, 57
Purines, 108
Pyrimidines, 108
Pyruvate dehydrogenase, 45

R

Racemic mixture, 4
Rate-limiting steps, 40
Receptor, 19, 96
Recombination repair, 116-117
Replication, 106, 112i, 113, 114i, 115
Respiratory chain, 48-49, 50i, 51-54
Restriction endonuclease, 137
Reverse transcriptase, 139
Ribonucleic acid (RNA), 106
  DNA and, 111
  translation and, 107, 120-122, 123i, 124, 125i-126i
Ribosome, 18, 19

S

Sarcomere, 65
Serine, 25, 26
Shine-Dalgarno sequence, 126
Side chains, 24
Signal sequence, 68, 127
Signal transducers, 93
SOS repair, 117
Southern blot, 138
Spectroscopy, 33, 35
Sphingolipids, 95
Sphingomyelin, 96i
Starch, 81
Stereochemistry, 1
Stereoisomer, 3, 6
Steroids, 95-96
Substrate concentration, 61i
Sucrose, 80i

i refers to an illustration; t, to a table.

T

Thioester bond, 38
Thiol group, 2i
Threonine, 25, 26
Transamination, 71-72
Transcription, 106, 117, 118i, 119-120
Transferase, 70
Translation, 107, 120-122, 123i, 124, 125i-126i
Triacylglycerols, 93-94, 95i
Tricarboxylic acid (TCA) cycle, 37, 45-46, 47i, 48
Tryptophan, 24, 25, 131-135
Turnover number, 63
Tyrosine, 25

U

Ultracentrifugation, 34
Urea, 2i
  cycle, 72, 73i

V

Vacuole, 20
Valine, 24, 25
van der Waals forces, 7
Vector, 139

W-Z

Water
  dissociation of, 9-10
  structure of, 8, 9i
Zwitterion, 27

i refers to an illustration; t, to a table.